Ian Slinton?
West Garth
Driffield
E. Yorks

SPACECRAFT
IN FACT AND FICTION

HARRY HARRISON AND MALCOLM EDWARDS
SPACECRAFT
IN FACT AND FICTION

ORBIS PUBLISHING
LONDON

Endpapers: A flying saucer lands on Altair-four in the film *Forbidden Planet* (1956).
Half-title page: 'Elopement Extraordinary, of Jack and his Lassie on a Matrimonial Excursion to the Moon in a New Aerial Machine'—an early nineteenth century lithograph.
Title page: Astronaut David Scott leaves the command module of Apollo 9 during rehearsals for the Moon landing.
Right: Testing 1932 style—first rocket tests by the American Interplanetary Society on a New Jersey farm.

© 1979 by S.F. Horizons Ltd.
First published in Great Britain by Orbis Publishing Ltd, London, 1979.

All rights reserved. No part of this publication may be reproduced, stored in a retrieval system, or transmitted in any form or by any means, electronic, mechanical, photocopying, recording or otherwise, without the prior permission of the publishers. Such permission, if granted, is subject to a fee depending on the nature of the use.
Printed in England by Jarrold & Sons Ltd, Norwich.
ISBN 0 85613 192 S

CONTENTS

FLYING TO THE MOON 7
THE FIRST ROCKETEERS 17
BEYOND THE SOLAR SYSTEM 31
FICTION BECOMES FACT 47
THE SPACE AGE 71
TRAVELLERS TO THE STARS 95
THE FUTURE OF THE FUTURE 115
BIBLIOGRAPHY 126
ACKNOWLEDGMENTS 126
INDEX 127

FLYING TO THE MOON

Left *A flaming chariot rather than rockets takes Orlando Furioso to the Moon in Ariosto's novel, illustrated in this late 19th-century edition by Gustav Doré.*
Above *Although Jules Verne knew there was a vacuum in outer space, the illustrator of* Round the Moon *has a somewhat breathless traveller waving his arms enthusiastically from a porthole.*

Until very recently the moon was visible but quite unapproachable. Very frustrating indeed. Dogs could howl at it, vampires flapped forth under its encouraging rays, while the Amerindians used its waning and waxing cycles as a calendar. But no one could get near it. It was very obviously an object of some kind up there in the sky—there never seems to have been any doubt about that. The tides in the sea followed its movement across the sky, the sun shadowed its surface and it could be seen to move in relation to the stars behind it. Oh, yes, it was there, tantalizingly close but still inaccessible.

If it could not be reached in reality, it could certainly be touched through fiction, the only object in the sky that the ancient civilizations found worth visiting. The sun was a ball of fire, that was obvious; after all, look what happened to Daedalus when he got too close. The planets were little wandering stars, while the stars themselves could be nothing other than lanterns hung from an arch or holes in crystal spheres. There was little interest in getting fried in a heavenly fire or crashing into a wall of crystal—but the moon, that was another thing altogether.

But how did one get to it? By flying of course, since it was up there in the sky, up in the air. The earliest known account of a flight like this, *Icaromenippus*, was written in the second century by a naturalized Greek Syrian named Loukianos of Samosata, or Lucian the Scoffer. Can this sneering Levantine be the father of all the science fiction flights to come? It is possible. Certainly his scientific preparations are as dim as many that were to follow; his hero flaps off to the moon in a highly unbalanced manner with an eagle wing strapped to one arm and a vulture wing strapped to the other. Earth's atmosphere extends to the moon, an error that has rung down through the ages meaning that all early spacecraft were envisaged as types of aircraft.

Lucian's first venture into primitive SF must have been a popular success, though the critics were probably contemptuous, since he followed it with another, *True History*. The birds' wings are laid aside this time and an entire ship is snatched up by a waterspout which carries it to the moon. The first of many sailing-ships of the clouds to come. And once on the moon there are great adventures with horse-vultures, cabbage-birds and giant fleas.

8 FLYING TO THE MOON

Heady stuff indeed, but it was obvious that the market was not ready for SF and spaceflight just yet. The Roman empire fell a few years later and the Dark Ages descended and people had other things to think about for a thousand years or more. No more moon flights were attempted until 1532 when Ludovico Ariosto, an Italian lawyer-turned-poet, wrote *Orlando Furioso (Mad Roland).* The journey to the moon this time is accomplished by borrowing Elijah's fiery chariot.

All of these moon voyages took place in the pre-scientific era when study of the natural sciences not only was not encouraged but was sternly frowned upon. The medieval astronomers worked with the threat of holy death hanging over them at all times. It was just eleven years after Ariosto's moon journey that Copernicus published his treatise on planetary motion. Kepler and Galileo followed up his lead, and the Earth was no longer seen as the centre of the

Left *In Cyrano's first attempt to reach the Moon the flasks of dew lift him swiftly and improbably into the air.*
Above *Rockets at last, the first of many to come, as Cyrano blasts off to the Moon—to reach burn-out earlier than planned. Luckily the marrow-powered second stage saves him from disaster: a crash landing among the palm trees of a singularly unrealistic Canada.*

universe. The Earth rotated about the sun and the planets were known to be other worlds like the Earth. But it still was not healthy to put these speculations into print. Kepler wrote a little scientific romance of his own, the *Somnium,* though it was not published until four years after his death in 1634, when he was well out of the reach of a vengeful church. Kepler was too much of a scientist to describe a flight to the moon—he was well aware that Earth's atmosphere did not extend that far, so he made the voyage in a dream. Instead of fanciful landscapes and creatures, Kepler attempted to describe the lunar landscape in the most scientific and accurate way.

The Protestant church was always a bit more scientifically inclined than the Catholic one. Even the bishops got into the spaceflight business; Bishop John Wilkins, in 1638, actually wrote a non-fiction book about moon-voyaging called *A Discourse Concerning a New World.* He even described the various options open for means of travel there:

There are four different ways in which flying in the air has been or may be attempted. 1. By spirits, or angels. 2. By the help of fowls. 3. By wings fastened immediately to the body. 4. By a flying chariot.

Not as scientifically knowledgeable as Kepler, the bishop still thought there was air all the way to the moon. But at least, among the helping fowls, he did mention a flying chariot: the first spacecraft ever to be described.

1638 must have been a good year for bishops, because Bishop Francis Godwin also published in this year: *The Man in the Moone.* He had a flying machine, but he kept the fowls as well, a species of super-swans he called 'gansas' which were tethered to the chariot for its lunar journey.

The first writer to hit upon the correct means of propulsion for spacecraft was the Frenchman Cyrano de Bergerac. His *Voyage to the Moone (Voyage dans la Lune),* written in 1649, is a wild tale meant mostly for entertainment, filled with jokes and puns, and was never intended to be taken seriously. Cyrano's first attempt to reach the moon is by dew-power, bottles of dew strapped to his body. Since dew 'rises' when the sun shines on it in the morning, the dew in the bottles is supposed to lift him into the air. Strangely enough it does, and Cyrano gets as far as Canada in this manner. Since the rising dew has let him down Cyrano next builds a spring-powered flying machine. He should have stayed with the dew, for this one crashes and knocks him about a good deal. He rubs bone marrow on his bruises to draw out the pain, then takes off again with firework rockets strapped to his machine. Success—at least for a while. The contraption soars high into the air until the rockets burn out. The thing crashes back to the ground but, most mysteriously, Cyrano keeps rising—all the way to the moon. The secret revealed at last; the moon has 'drawn' the marrow on his body towards it just as the marrow 'draws' out pain.

Man was not meant to fly by puns alone, so it is nice to see between the 'rising' dew and the 'drawn' marrow a good, solid, black powder rocket. Tongue-in-cheek as the author is, he at least made one correct guess, and it took over 200 years for any other

Rising up from Ascension Island, Domingo Gonsales, hero of Bishop Godwin's The Man in the Moone, *idly steers his ten birdpower construction with the aid of a movable sail.*

Above *With a clap of thunder and a mighty burst of flame the manned projectile in Verne's* From the Earth to the Moon *is shot into space from the world's largest cannon.*

Top *Verne's travellers experience the delights of a zero gravity ballet. The return trip is not quite to entertaining as the space capsule* (Right) *plummets back to Earth.*

author to latch on to the rocket as a valid means of propulsion for spacecraft. They continued to follow Bishop Wilkins's four dicta right up until the end of the nineteenth century. Spirits continued to lift and fowls to tow.

Nor was there any attempt ever made to equip the vehicles for the rigours of a space voyage. The cold of space, the lack of air, chance meteorites and other inconveniences were disregarded in 'scientific romances' which were obviously more romantic than scientific. The first reasonable spaceship design was made by an American, Professor George Tucker in 1827, who hid his academic credits behind the pen-name of Joseph Atterley, in his book *A Voyage to the Moon.* 'Atterley' uses the familiar device of that time of presenting the story as a real adventure. In his travels he meets a Brahmin who has discovered a new metal which he calls Lunarium because of its unusual properties. It is repelled by the Earth and attracted by the moon. This gives Atterley an interesting idea—which the reader will already have guessed if he has glanced at the title of the book—so with the Brahmin's aid he builds a cubical copper vehicle, well insulated and complete with an air supply. They secure the thing to the ground with thick ropes before coating it all over with Lunarium. They climb into the vehicle and cut the ropes, seal the hatch—and after a few not-too-brief speeches step out on to the moon three days later—where they have the traditional adventures with strange creatures.

The importance of this book is that Tucker was actually aware of the realities of spaceflight and made some attempt to find realistic solutions to the problems. The vehicle is a recognizable spaceship—it is insulated, sealed and supplied with air. There are also no spirits or fowls to lift and tow the thing, but instead a miraculous invention of anti-gravity metal provides propulsion. This was the first use of an anit-gravity device to move a spacecraft, an invention that was to lift many a vehicle in nineteenth-century science fiction. All of these romances are justifiably forgotten except for the most able of them, H. G. Wells's *First Men in the Moon.* Tucker's book may also have been responsible for the first burlesque of space adventures, 'The Unparalleled Adventures of One Hans Pfaall' by Edgar Allan Poe. Tucker was a professor at the University of Virginia where Poe studied. Poe had certainly read the professor's story by the time he wrote his parody in 1835. Although his lunar travellers make the voyage in a balloon, the rest of the scientific descriptions are very accurate for the time.

Could Poe's story also be the genesis of the incestuous nature of science fiction? A custom that has continued to the present day. Perhaps. Certainly Jules Verne was a great admirer of Poe, even producing a sequel to Poe's unfinished *Narrative of Arthur Gordon Pym.* Verne's 1865 novel, *From the Earth to the Moon,* not only tips its hat to Hans Pfaall with little bits of reference, but it also attempts to be facetious in a leaden Franco-Teutonic manner Verne was very proud of this novel and of its scientific accuracy, which is a good measure of Verne's almost total lack of knowledge of science. In the book the Baltimore Gun Club build a spectacularly large cannon with a barrel 900 feet long. Since it would be a little hard to construct a carriage for a weapon this big it is sunk vertically into the ground instead. It is loaded with 200 tons of guncotton on top of which is placed a

12 FLYING TO THE MOON

Left *In 1902 George Melies'* A Trip to the Moon *launched the first space capsule on film—hitting the Man in the Moon squarely in the eye.*
Right *Martian airships collide in Griffith's* A Honeymoon in Space.
Below *Railway buffers soften the impact of landing in the film version of* First Men in the Moon.

projectile nine feet in diameter with walls a foot thick. Originally this was to be a great roundshot, a sort of super-cannonball, but its shape is changed when three passengers agree to be fired off in it. When it is finally lowered into place on the lethal tons of guncotton it has the pointed nose and proportions of a traditional artillery shell.

The gun is fired and off the intrepid travellers speed. The concussion of the firing knocks them out, but they recover in time to feel the heat of friction of the projectile's passage through the atmosphere, followed by the chill of space. Later they float about and enjoy the effects of zero gravity at the *bouleversement,* the point at which the gravitational attraction of the Earth and moon balance exactly. This interesting phenomenon was previously described by Poe, Tucker and others, despite the fact that it does not exist. The craft reaches the moon about 97 hours after launching, whips around it and thence back to Earth for a happy landing.

This is the kind of pseudo-science that gives science fiction a bad name. Everything, including Verne's elaborate calculations, is completely wrong. He got his sums wrong by a factor of ten. If his preposterous gun did not simply blow up upon firing, the shell would have emerged with a velocity of approximately 1,200 yards per second. Not the 12,000 yards per second Verne mentions as needed to escape from Earth's gravity. This speed would have sent the thing about 12 miles straight up into the air to drop back to Earth with a resounding clang.

Not that this would have bothered the passengers inside. If they were unlucky enough to be subjected to sudden acceleration of this kind—from zero to 24,545 miles an hour in 500 feet if we use Verne's figures—they would all be mixed into a sort of red jelly on the back wall of the ship. At this speed of acceleration, a 170-pound man would weigh 3,422 tons.

The closest Verne got to a touch of real science was the rockets he attached to the shell to break their fall when they land on the moon. However, they are not used in the story, so their ridiculously small thrust cannot be laughed at. Instead, the craft is

pulled out of correct orbit by the gravitational pull of a passing meteor; just as impossible as the rest of the science in the novel.

One writer who did stick with the rockets was another Frenchman, Achille Eyraud, whose *Voyage à Venus* was published in the same year as Verne's *From the Earth to the Moon*. His spaceship is propelled by a 'reaction motor' or rocket, and it makes its trip to Venus in this fashion. But what the novel gained in scientific accuracy, it lost in readability and therefore fell deservedly into obscurity while Verne's still marches resolutely into the future.

The rest of the nineteenth century is filled with stories of space travel, almost all of them deservedly sunk into oblivion. Only a few are of interest to the modern reader. Edward Everett Hale—a crusading clergyman who is usually known by a single story, 'The Man without a Country' (such is fame; he published over sixty books)—wrote *The Brick Moon* in 1870. This contains the first proposal for an artificial satellite to be launched into orbit from the Earth, as a most sensible aid to navigation. While the idea is a sound one the science, alas, is not. The 200-foot thick brick satellite is to be hurled upwards by two gigantic flywheels spinning in opposite directions. The preposterous thing is launched, by accident, and is flipped into orbit with the brickies and their families still inside. They survive nicely in orbit, supplied by food slung up by the flywheels.

By 1880 things in space were getting a good deal better. Percy Greg's great long two-volume novel *Across the Zodiac* describes the voyage and the ship with attention to scientific detail. The spacecraft is made of airtight metal, insulated with cement and supplied with portholes to look out at the interplanetary scenery. It is propelled by 'apergy', a new electric force that counters gravity but which is not explained in too great detail. The narrator travels to Mars in the device, landing there safely to discover an advanced human civilization. His return trip to Earth, however, is not quite as successful. What is later taken to be a new meteor crater is discovered to be the hole made by his ship crashing at speed; this is revealed by his journal found in a metal box in the crater.

Anti-gravity propulsion was a standby for SF writers of the Victorian era, this and the giant cannon. H. G. Wells must have been most suspicious of the cannon-powered flight for he used it only once in a novel, in *The War of the Worlds* published in 1898. Even here it is fired off-scene to launch the Martian invaders to their decayed doom on Earth. When he wished to show spaceflight a little closer up, in *The First Men in the Moon* in 1901, he fell back on the old anti-gravity metal. His inventor, Cavor, deduces that gravity is a form of radiant energy, just like light. Since there are substances that block light waves there must also be a material that will block gravity waves. After a bit of experimentation he finds just such a substance. Not being as humble as the Brahmin who named his metal Lunarium, Cavor calls his Cavorite. Though Wells fudges his science with Cavorite, he designs his spaceship with far more ingenuity. It is well equipped with food, water and air, and built in the shape of a sphere. The outer hull is covered with sections of Cavorite, each of which can be rolled up like a blind. The blinds beneath are opened out and the thing shoots up into the air. As it approaches the moon the blinds on that side are rolled up, exposing the vehicle to the moon's gravitational field. Thus the vehicle can be steered by adroit

Top *A 1911 German spacecraft serenely approaches the Moon in F. W. Maler's* Wunderwelten.
Bottom In The World in 2030 *the air traveller enjoys more leg room than today.*
Right *In Fenton Ash's* Son of the Stars *the Martian landscape is far more hospitable than the one we know, and the spacecraft suggest imperial grandeur.*

opening and closing of the blinds; good exercise too. The voyage does end on the moon, where Cavor and his travelling companion discover the underground civilization of the Selenites. Their society, one of Wells's dramatizations of Darwin's theories, is based on a community like that of the bees or ants, where all individuals have highly specialized roles.

The last quarter of the nineteenth century saw few other interesting or important spaceflights in literature. The new century, however, began with a new anti-gravity power called R Force. In 1901 this powered the spacecraft that was both vehicle and nuptial bed for *A Honeymoon in Space* by George Griffith. The extrovert Lord Redgrave, having financed the scientists who discover R Force, uses it to launch the spacecraft for this rather unusual purpose. In addition, we assume, to embracing his bride, the lord's grand tour embraces most of the major planets and a few of their satellites.

More important, the turn of the century saw the birth and growth of the pulp magazines in America. These cheaply printed and garish journals, named for the paper they were printed on, were designed as simple-minded entertainment for the undiscerning mass market. A popular pulp author was Garrett P. Serviss, who first became known in 1898 for his newspaper serial, *Edison's Conquest of Mars*. This is a sequel to Wells's *War of the Worlds* in which the bold scientist of the title lays aside the inventing of the light bulb for a bit in order to lead his fellow Earthmen in a vengeful attack on the Martian's home planet. It is obvious that Serviss could read as well as write, because in 1901 in *A Columbus of Space* his spacecraft gets to Venus propelled by atomic power—derived in fact from uranium.

Other interplanetary romances were not as scientific. Before embarking upon his Tarzan series, Edgar Rice Burroughs published *Under the Moons of Mars* in 1912. This is still in print as *A Princess of Mars,* and is noted not only for its egg-laying Red Martians with belly-buttons, but also for its sidestepping of the transportation problem. The hero John Carter, who later marries the princess of the title, arrives on the planet by what appears to be astral projection.

Burroughs was not the only one who had trouble getting to Mars. The Earl of Birkenhead got into the act in 1930 with his *The World in 2030 A.D.* Notable barrister and politician that he was, he was a little short on his science. In it he wrote:

By 2030 the first preparations for the first attempt to reach Mars may perhaps be under consideration. The hardy individuals who form the personnel of the expedition will be equipped with a number of light masts which can be quickly extended like fishing rods, from its nose. The purpose of these will be to break the impact with which, granted all possible skill and luck, the projectile would strike the surface of the planet.

What a lovely image the noble lord's fishing-pole equipped rocket does make! More interesting things were happening in Russia. In 1916 the journal *Priroda i lyudi* published part of a very realistic novel about space travel. Unhappily the magazine folded after printing only half of the novel. Although it had been written in 1896, the entire book was not to see print until 1920. It had been written by the then unknown scientist Konstantin Tsiolkovsky, and was called *Beyond the Planet Earth*.

THE FIRST ROCKETEERS

Rockets have been around for a long time. In the year 1232 the Chinese lifted the siege of Kai-fung-fu by dropping bombs off the city walls on to the Mongols' heads—then shooting them with 'arrows of flying fire'. The contemporary description of the arrows leaves no doubt that they were rockets. Like any good weapon their use spread around the world very fast, so by 1258 they were being used in Cologne and in 1379 *rocchette* are credited with winning the battle for the Isle of Chioggia, near Venice.

The history of rockets shows them alternating between being used as fireworks or as weapons. Primitive cannon seemed to work a good deal better than primitive rockets so their principal use seems to have been in fireworks displays until the nineteenth century in Europe, although the Indians were using them to blow up the British in the eighteenth century. The invading troops found themselves bombarded with iron-tubed rockets weighing up to 12 pounds, guided by 10-foot long bamboo poles. Their range was a mile and a half and they were not very accurate, but could still inflict damage on massed troops. The British did nothing about investigating this novel weapon for over twenty years. It was then that a certain Colonel William Congreve read about the Indian rockets and decided to do some experimenting on his own.

At first Congreve used store-bought models to experiment with, but soon started manufacturing on his own. By lucky chance his father was a general and in charge of the Royal Laboratory at Woolwich, which provided not only gunpowder but military attention for the project. Congreve developed a range of military rockets, the biggest weighing 42 pounds with a range of almost two miles. With these weapons the British first burnt and then captured the cities of Boulogne, Copenhagen and Danzig during the Napoleonic wars. All of the world armies then formed rocket companies, but abandoned them as the military cannon improved far beyond the rocket. It was back to the fireworks once again.

The lifting ability of rockets, for anything other than military incendiaries, was demonstrated a few years later by Claude Ruggieri, an Italian living in Paris. He delighted the locals by blasting small animals up into the air on rockets, whereafter they dropped to safety in parachutes. The high point of his demonstrations was reached in 1830 when he planned to send a ram into the air on a

Left *In 1931 Karl Poggensee launched a solid fuel rocket which reached a height of 1500 feet before dropping back to the launch site near Berlin.*
Inset *Since rockets can blow up rather than burn up, Poggensee and assistants watch the launch from the questionable safety of a barn.*

monster rocket. When a human volunteer offered to take the sheep's place the police closed the entire operation down, since the volunteer was in fact a small boy.

The boy would have been the first human rocketeer, unless we believe the story of Wan-Hoo, a Chinese official who built a flying machine in 1500. This was made of two kites fastened together and powered by a number of rockets. When the fuse was lit, however, the whole thing blew up and no identifiable piece of either inventor or invention was ever found.

The later application of rockets to spaceflight seems to have been developed at the same time by three men, working separately in different countries. It has been said that when it is steamship time, steamships will be invented. The beginning of this century was apparently rocket time. The three were Konstantin Tsiolkovsky, a Russian, Robert H. Goddard in America, and Hermann Oberth, a German.

Chronologically Tsiolkovsky was the first (1857–1935) so credit must be given to the Soviet claim that he is the 'Father of Space Travel'. He was born in the tiny and remote village of Ijvesk and seems to have been largely self-taught. He was so far removed from

Above *A Congreve war rocket is launched with great aplomb by a battery of the Royal Horse Artillery in 1835.*
Above right *Ganswindt's interesting but highly impractical rocket ship.*

the mainstream of science that he spent many years rediscovering various scientific principles already well-known in the outside world. When he submitted various of these discoveries to the St Petersburg Society for Physics and Chemistry he was at first thought to be a madman or a charlatan. But when it was realized that he had done this remarkable work on his own he was very quickly admitted to membership, sharing this glory with scientific eminences like Mendeleyev, the discoverer of the periodic table of elements, who sponsored him.

Since the age of fourteen Tsiolkovsky had been interested in flying machines, and around 1878 he widened this interest to take in the theory of spaceflight as well. He worked on the problem for a number of years, but unhappily did not publish anything on the subject until 1895. By this time a slightly eccentric German inventor by the name of Hermann Ganswindt had begun to publicize his idea for a spaceship by a novel form of rocket. Ganswindt was long on theory but more than a little short on knowledge of physics. He never did grasp exactly what Newton was about with his Third Law of Motion. Nor is he alone in this; people still ask how rockets can fly through space when there is nothing to push against. Even as august a journal as the *New York Times* got it wrong when they published this in 1920:

That Professor Goddard, with his chair in Clark College and the countenancing of the Smithsonian Institute, does not know the relation of action to reaction, and the need to have something better than a vacuum against which to react—to say that would be absurd. Of course he only seems to lack the knowledge ladled out daily in high schools.

It is of course the editorial writer who lacks this elementary knowledge.

The famous Third Law causes a lot of confusion—yet it should not. Anyone who has fired a rifle and got banged in the shoulder has experienced the statement that for every action there is an equal and opposite reaction. The same amount of energy that throws the light bullet for a mile moves the heavy gun back a few inches. So what has this got to do with rockets in space? Everything. It all has to do with pressure in a sealed container.

If you take a hollow metal sphere and explode some gunpowder in it the pressure on every square inch of the inside of the sphere's surface will be the same. Make the sphere into a tube with sealed ends and the pressure is still the same anywhere on the interior surface. Now make the tube into a rocket by taking off the bottom end. It flies up into the air. Why? Look at diagram A.

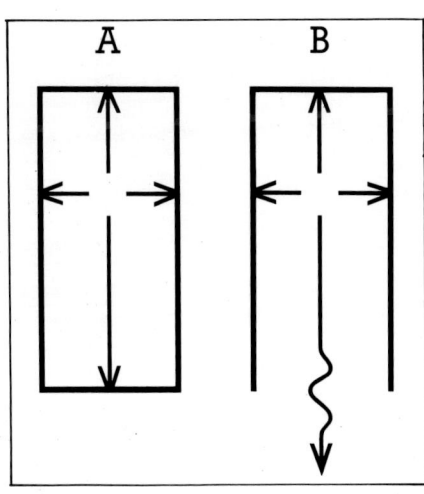

The pressure inside is, let us say, ten pounds. So the pressure on the side walls of the tube is ten pounds—pushing out sideways in all directions. Equal push means no push, so the tube does not move sideways.

But, since the bottom end of diagram B is gone, there is no pressure here. There is nothing to equalize the ten pounds' pressure on the top end, so it is pushed up with enough power to lift ten pounds. The can goes flying up into the air with ten pounds of 'lift' Lift is now a technical term meaning just that. Ten tons of lift will lift ten tons.

20 THE FIRST ROCKETEERS

To get lift in a rocket we throw away gas very fast. Just like throwing away a small bullet very fast. The bigger the bullet and the faster it is thrown, the more bang on the shoulder. The more gas and the faster it is thrown, the more lift. Since the speed that the gas is thrown away at is an important factor in lift, it can easily be seen that the burning gas does not have to push against anything to get thrust. In fact in space where there is nothing to slow it down the rocket works that much better. Quite the opposite of the accepted worry.

If the *New York Times* could get it wrong, and most laymen be confused by it, why then it is small wonder that poor old Hermann Ganswindt never did quite grasp what it was all about. This law student-turned-inventor intuitively grasped the fact that reaction could be used as a means of propulsion in space. But he could not believe that a gas jet could produce sufficient reaction to lift a solid body. So he came up with the idea of steel cartridges filled with dynamite that exploded out lumps of steel. A passenger cabin suspended below this engine would have an open shaft built through it for these flying chunks. Happily this unusual craft was never built; had it been there might have been a shortage of volunteers to sit in the cabin under the exploding dynamite with the shrapnel whistling by just beside them.

Tsiolkovsky, self-trained scientist, did much better. He realized that solid fuel explosives available in the nineteenth century just did not have enough power to lift a spacecraft. Liquid fuels such as kerosene would be much better, he believed, and in 1903 he published a design of a spaceship whose propellant was liquid hydrogen and liquid oxygen. That he was on the right track is

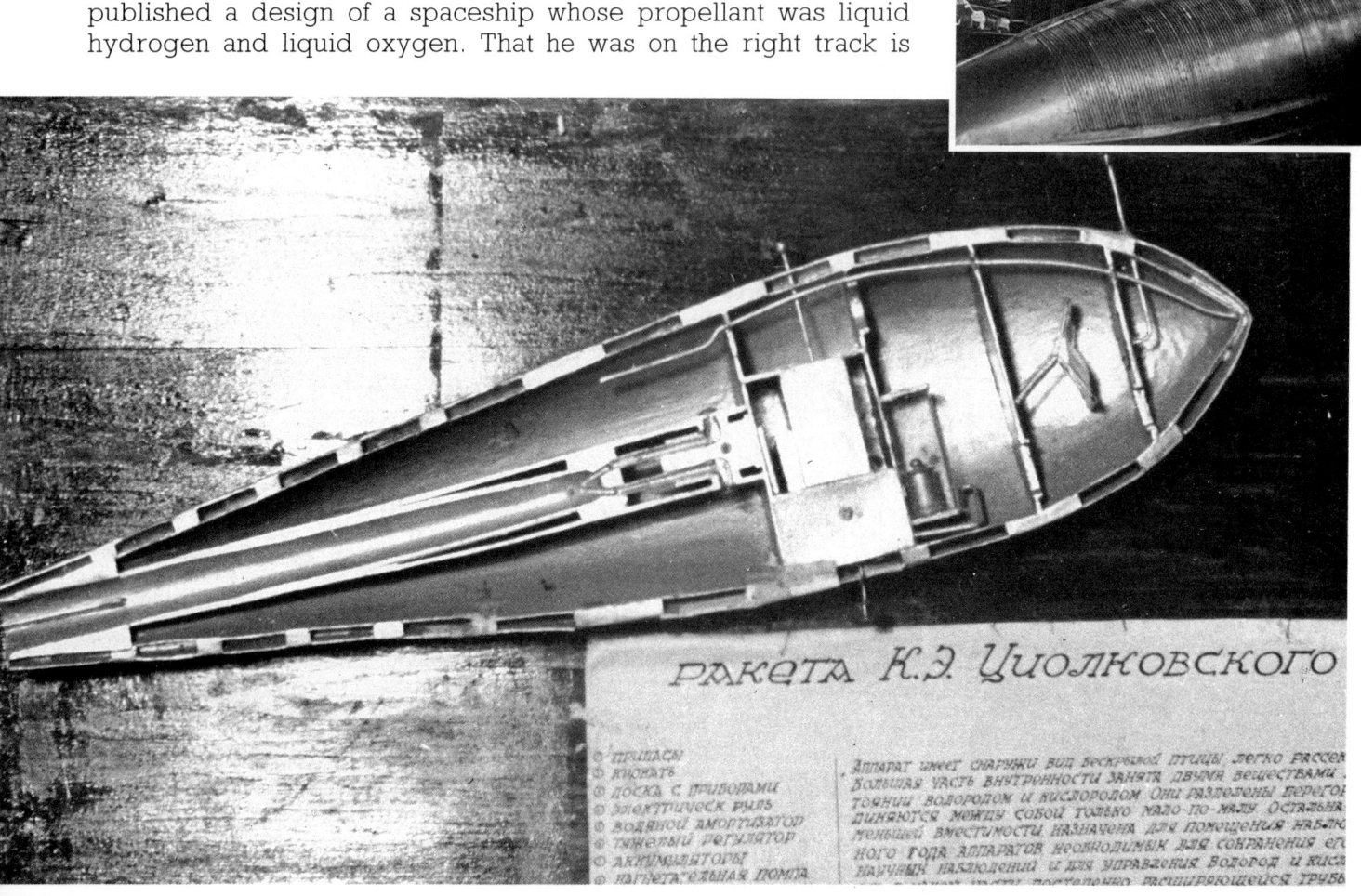

Below *Konstantin Tsiolkovsky with a model of one of his rockets.* Bottom *A cutaway showing the rocket's interior, with details of the firing chamber.*

obvious now that we know that the American Saturn rockets had a first and second stage fuelled by kerosene and oxygen, while the third stage used hydrogen and oxygen, all liquefied. Multi-stage rockets of this type were also proposed by Tsiolkovsky. He saw the advantages of discarding a fuel tank when it was empty, thus leaving less mass to be lifted by the remaining stages.

Almost all of Tsiolkovsky's ideas are worked out in his novel *Vne Zemli (Beyond the Planet Earth)*. He began writing the book in 1896, but it was not published until 1920. The English edition had to wait until 1960. As a work of fiction it is no winner, being overly long and more given to serious lectures than action. What plot it has is just there as a structure for the author to hang his revolutionary ideas upon. The novel describes the construction and operation of a spaceship by an impressively international group of scientists. There is of course a Russian, as well as a Frenchman, a German, an Englishman, an Italian and an American. The Russian is called Ivanov, the equivalent of Smith or Jones, but the others are far more impressively named: Laplace, Helmholtz, Newton, Galileo and Franklin. With a brains trust like that the rocket should get off the ground. After much discussion of technical problems and the testing of prototypes, the final design is begun:

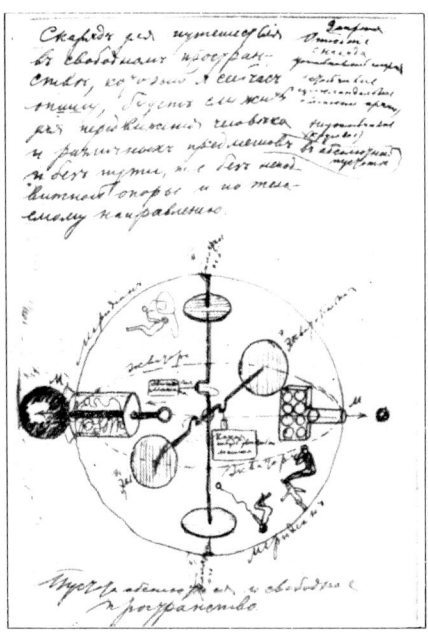

Below *Tsiolkovsky's sketch of a jet-propelled spaceship, showing passengers in free fall.*
Bottom *His design for a space greenhouse.*

The composite model consisted of a streamlined body a hundred metres long and four metres wide, rather like a gigantic spindle. It was divided by transverse partitions into twenty compartments, each of which was in itself a jet apparatus; that is, each compartment contained a fuel supply and constituted an explosion chamber with an automatic injector, a blast tube and so on. Only the central chamber had no jet equipment and served as a wardroom, 20 metres long and 4 metres in diameter ... The ingenuity of a rocket's design gave it comparatively light weight combined with very great efficient lift.

Because Franklin, in his canny new-world manner, has invented some new fuels this does not have to be a multi-stage rocket. Instead its twenty chambers, or engines, work in concert, fed with fuel by the injector control system. The other details of the construction of the craft are also explained in some detail. The outer skin is made of three separate layers of metal with cooling gas circulating between them. Even though the outer layer becomes white-hot during liftoff, little of this heat penetrates to the interior. Great care is taken to protect the passengers from the rigours of space travel. Instead of acceleration couches they are submerged in liquid baths, flying amniotic wombs. On liftoff the ship weighs 320 tons, three-quarters of which is fuel. The living quarters occupy over 14,000 cubic feet of space, equivalent to the area of a fair-sized house.

It is obvious when we compare his design with a modern ship that Tsiolkovsky knew what he was about. The Saturn V Skylab launcher stood 358 feet high and put into orbit a payload of 86 tons that had a working volume of nearly 13,000 cubic feet. Tsiolkovsky's rocket was 328 feet high and its payload was a little over 14,000 cubic feet. The main difference between the two ships is the amount of fuel carried. Franklin, the American, invented a fuel with more power in it than anything his countrymen used half a century later. Tsiolkovsky's ship gets into orbit after burning only 240 tons of fuel. Since the first stage of the Saturn V burns over 13 tons of fuel a

22 THE FIRST ROCKETEERS

Right *Goddard in 1926 with an early version of his liquid fuelled rocket. The Saturn V launchers are direct descendants of this crude construction.*
Below *Goddard proudly displays the apparatus which in 1916 proved his theory that rockets worked more efficiently in a vacuum.*

second, the overall weight of the craft at take-off was over 2,700 tons. Eight times heavier than the Russian ship.

Once they are in space, the international crew go right to work. As one of the characters explains:

'You, of course, gentlemen,' he went on, 'know how vast and free is the space that surrounds our Earth; you know that it is filled with light; you know that it is empty. It's a sad thought that we are crowded on Earth, treasuring every sunny corner where we can raise crops and build our homes and live in peace and tranquillity. While I was wandering the emptiness about our rocket, it was the vastness, the freedom and lightness of movement that most impressed me—that tremendous quantity of solar energy going to waste, uselessly. Who is there to stop men from building their greenhouses and their palaces here, and living in peace and plenty?'

Nothing is up there to stop them and they get right to work, at least on the greenhouses if not on the palaces. They rapidly construct an idyllic and self-sufficient environment, growing their own food and generating oxygen in their greenhouses, while the limitless solar energy supplies all the power they need. Not only has Tsiolkovsky described the first practical spaceship, but he has designed the first space colony as well. His pioneers are followed by thousands of volunteers who construct a giant network of colonies at a height of over 20,000 miles. At this height the colonies will orbit the Earth in exactly one day. This geosynchronous orbit means that they will rotate just as fast as the world below them, therefore they will appear to stay fixed over the same point on the surface below. If Tsiolkovsky had mentioned the telecommunication advantages of this kind of orbit he would have anticipated Arthur C. Clarke's 1945 proposal for TV satellites by a number of years. But his space-pioneers have plenty of other things to do, establishing more colonies between the orbits of Earth and Mars, and later in the asteroid belt to take advantage of the raw material easily available there.

Beyond the Planet Earth is the fictionalized representation of the ideas of a most visionary scientist. Tsiolkovsky is impatient for mankind to free itself from the constraints of an earthbound existence and he outlines the manner in which this can be done. But he kept his more imaginative speculation confined to the pages of his novel; his scientific papers continued only to explore the day-to-day problems of astronautics. As an abstract scientist he survived the 1917 revolution and became a member of the Soviet scientific hierarchy, writing and publishing prolifically right up to his death in 1935. But he was primarily a theorist and never performed the experiments that would have confirmed his ideas. Perhaps that was one reason why his pioneering work on rockets was not capitalized upon in the Soviet Union.

The second major figure, the American Robert Goddard, was just as much an experimenter as a theoretician. He was born in 1882 and his interest in space travel was sparked off as a teenager by reading *War of the Worlds* and *Edison's Conquest of Mars*. Completely on his own he decided that rockets were the only feasible means of propulsion in space, then developed the idea to include multi-stage vehicles. As he theorized so he began systematic

Below *The Saturn V crawler weighs 2,750 tons and transports the rocket to the launch pad at 2 mph—in Goddard's day the equipment was less complicated.*

experimentation, first testing his rocket motors in a vacuum chamber to prove that they did function more efficiently in airless conditions, then test-firing them in the field. By the time he was sending rockets to the height of 500 feet he applied to the Smithsonian Institute for a grant, and received one in 1916. By then he was a professor of physics at Clark University in Massachusetts and ready to publish. In 1919, to justify his grant, he wrote a report on his work that was brought out by the Smithsonian.

This was a very serious paper titled *A Method of Attaining Extreme Altitudes,* and was mostly concerned with the ability of rockets to carry instruments into the upper atmosphere to make observations there. However, he did mention in passing that with the development of rockets there was the possibility of someday building one that would reach the moon, where it would explode a small charge. This would be seen from Earth by telescope and would signal that the flight had been successful. The important scientific nature of Goddard's book was completely ignored as the newspapers seized on the moonshot and ridiculed Goddard as a crank. He was immensely annoyed by this, so much so that he published as little as possible after that about the results of his rocket experiments. However, he did attract the attention of Charles Lindbergh who arranged for him to receive a $50,000 grant from the Guggenheim Foundation in 1930. With this money Goddard moved his testing facilities from Massachusetts to the wide open spaces of New Mexico.

Like Tsiolkovsky he decided that liquid propellants would be needed for his rockets. In March 1926 he launched the first rocket of this kind. It rose to the majestic height of 41 feet and stayed up for all of two and a half seconds. This was just the beginning. By 1935 his liquid rockets were hitting an altitude of 7,500 feet. When World War II began, Goddard tried in vain to interest the U.S. government in the possible military uses of rockets. Eventually he did convince the Air Corps of the ability of rockets to assist heavily laden planes on take-off, and developed the JATO (Jet-Assisted Take-Off). He was still working on this project when he died in 1945.

The public knew him as serious physicist, a respectable researcher who laboured hard to improve the design and efficiency

Above left *The four rocket motors and tail piece of Professor Goddard's rocket, November 1936.*
Above *By 1937 Goddard had developed gimballed steering which assured a vertical liftoff.*

THE FIRST ROCKETEERS

of the rockets he had first conceived. Having been once bitten by the press he kept any more imaginative speculations to himself. Only after his death did his diaries reveal that he had been as advanced as Tsiolkovsky in his thinking, and was considering the possibility of interstellar travel as early as 1920.

Hermann Oberth was the third of the rocket pioneers. Born in 1894 in Transylvania, then a part of Hungary and famous as well for its vampires, he later moved to Germany, and became a German citizen. Like Goddard he had been inspired by early space fiction, the works of Jules Verne which he read as a child. Just before World War I he began writing about rockets, and in 1917 proposed a missile fuelled with liquid propellants. He did not read Russian or English, and had never heard of Tsiolkovsky or Goddard. It was liquid fuel time, and the three men developed the same idea completely independently.

Unlike his Russian and American contemporaries, Oberth did not mind putting his ideas into print. In 1923 he published a detailed technical study titled *The Rocket into Interplanetary Space.* It

Right *By 1940 Goddard's rockets included many improvements, including pump turbine units for propellant injection.*

included plans for a rocket called Modell B. This was a tapering cylinder mounted on fins, clearly the prototype design for the German V-2 rocket, as well as uncounted science fiction illustrations. This design is still preserved in solid metal in the Hugo, the annual science fiction writing award.

Oberth's book was received with great enthusiasm and interest. In 1927 a society for the promotion of space travel was founded in Breslau. This was the *Verein für Raumschiffart,* or VfR, which published the world's first rocket magazine, *Die Rakete.*

The next venture into space was made in 1928 by the great German film director Fritz Lang. He had already made one science fiction film, the classic *Metropolis,* which had been written by his wife, Thea von Harbou. Inspired by Oberth's book, and another popular volume on space travel by Willy Ley, she began working on the script for a film about space travel. Oberth himself was hired as technical consultant and proceeded to design a large and plausible spaceship for the film, *Frau im Mond* (known in English as *The Girl in the Moon* and *By Rocket to the Moon*). The film company also put up the money for Oberth to come to Berlin from Transylvania and build a real liquid-fuelled rocket that would be launched as a publicity stunt just before the film's première. This proved not to be as easy as planned. There was a ridiculously short time allowed to

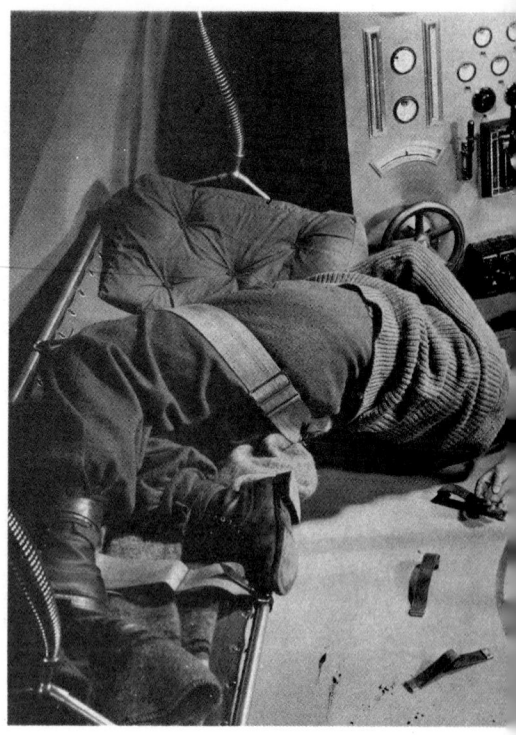

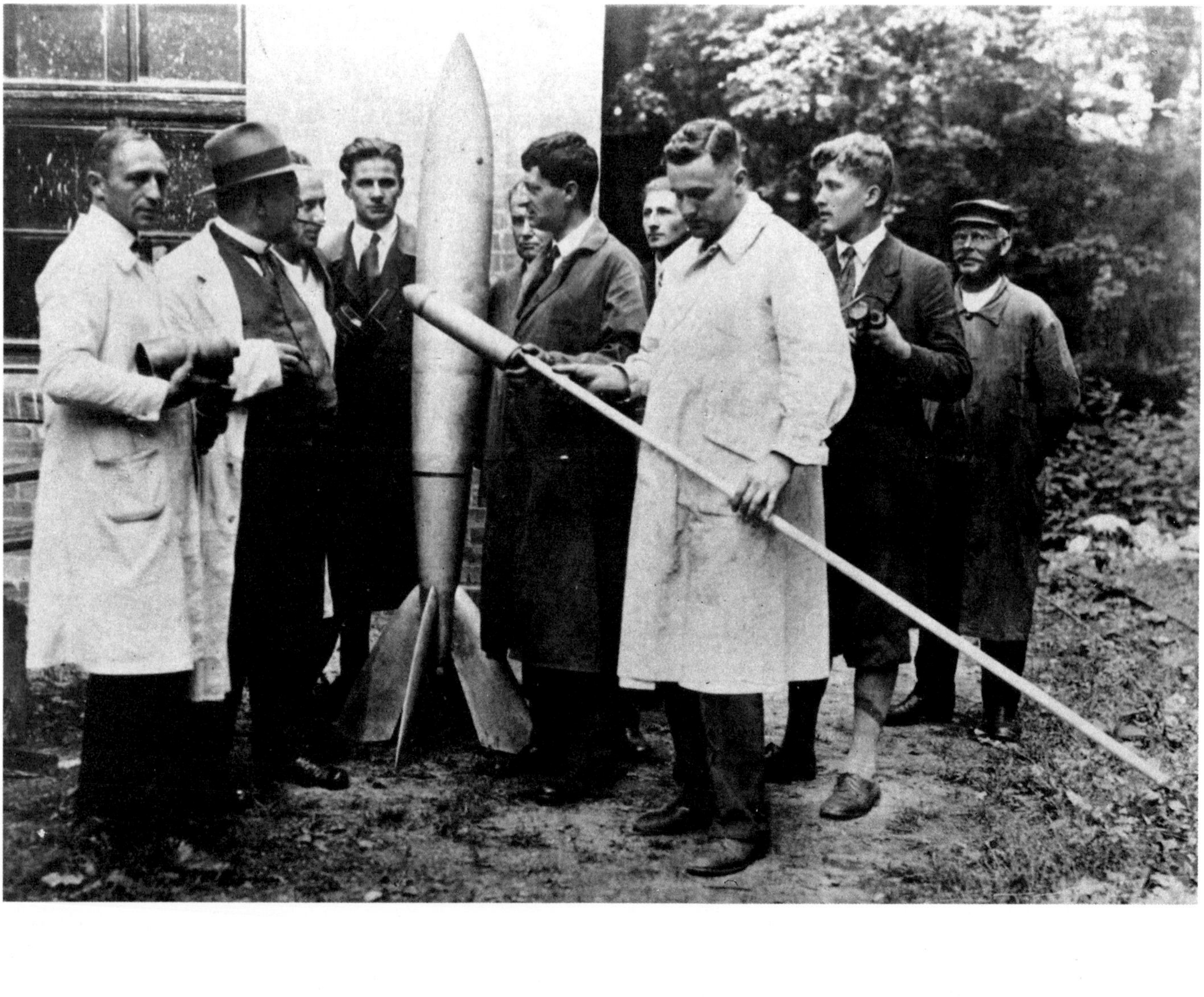

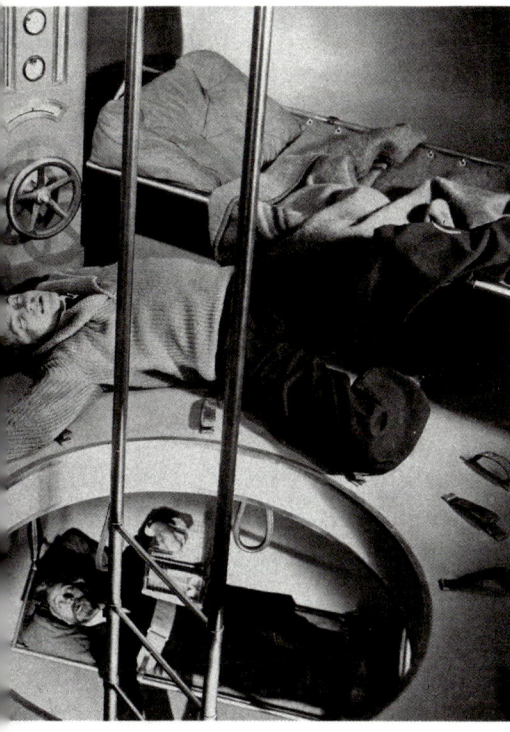

Left *A group of German rocket pioneers. Hermann Oberth is the figure immediately to the right of the rocket, which is one of his designs.*
Above *The passengers in* Frau im Mond *are left unconscious by the stress of liftoff.*
Right *The book of the film. Note the rocket fins apparently cast from concrete.*

design and build it in, just over twelve weeks. Oberth was a theorist and no engineer, basically a small-town schoolteacher, and unable to work under pressure. He hired incompetents to help him and the project was beset with one problem after another; a good launch site could not be found, an explosion nearly cost Oberth the sight of one eye, and the original design had to be abandoned to construct a simplified demonstration model. Oberth was annoyed and confused by the rush and bustle, by what he called 'the behaviour of these soulless, money-making, German-speaking Americans who call themselves Berliners'. Though the rocket was never launched the film was and became an instant success. Oberth packed up and went

JOURNAL

OF THE

British Interplanetary Society

JANUARY 1939 6d. to non-members.

Design for a Lunar Space-ship. See article page 4.

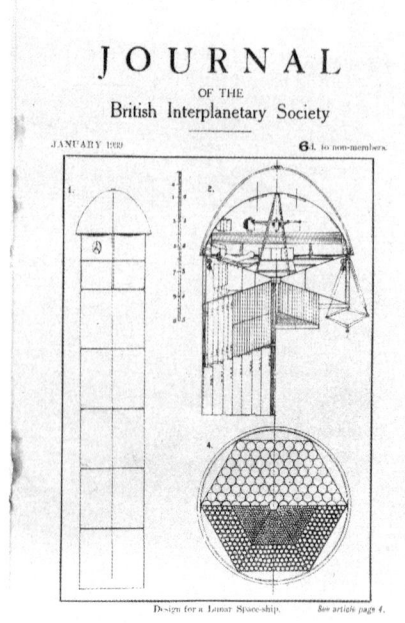

Above *The British Interplanetary Society's design for a potentially explosive six-stage Moon rocket.*
Right *On the pad at Peenemunde in 1942 a V-2 is prepared for launching.*
Far right *A successful launch—one of the few at that time.*

back to Transylvania muttering threats of lawsuit. Later he blamed his erratic behaviour on the explosion, which gave him shellshock.

But the VfR, now five hundred members strong, would not be stopped. In 1929 they elected Oberth their president, and Willy Ley vice-president. The film company turned Oberth's rocket over to them. Upon examination everyone—including Oberth—agreed it would not work. Another design of Oberth's was built instead and in 1930 the VfR conducted what they thought was the first successful test of a liquid-fuelled engine. They had no idea that Goddard had sent his aloft in 1926 since he refused all attempts at communication. Having seen his idea launched in reality, Oberth was more than happy to return to his high school teaching in Transylvania. He was the only one of the three rocket pioneers who lived to see the dawning of the space age.

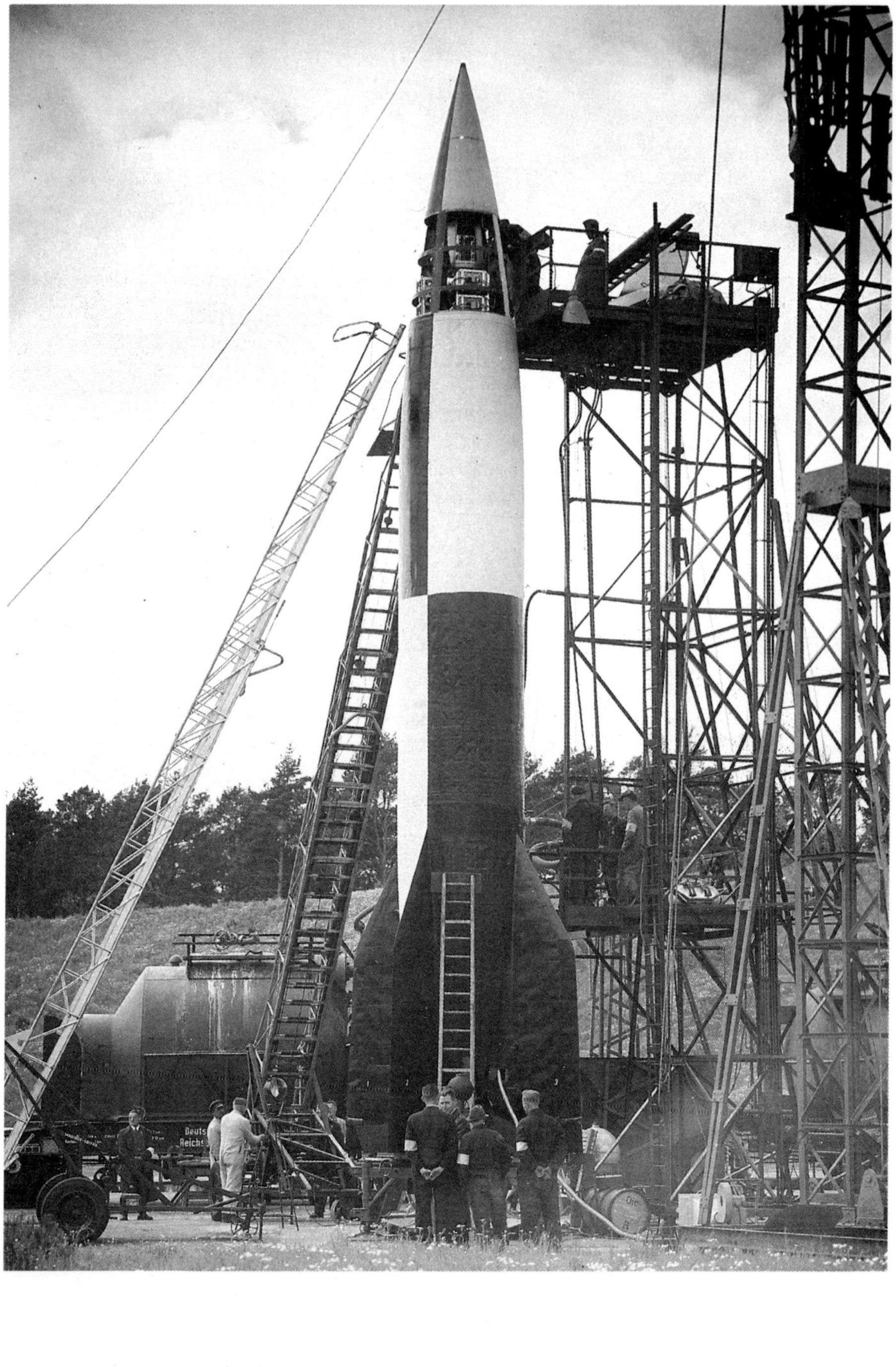

The VfR continued its testing programme, and in 1931 fired off its first successful rocket. However, the depression that was rocking Germany had a deadly effect on the VfR's membership and finances, and by the end of 1933 their experiments had ceased. But their spirit went marching on in the form of the young Wernher von Braun. The German army had been watching the experiments with an appreciative eye. They hired von Braun as a civilian attached to their rocket development programme, which ended up at Peenemünde. He was not the first scientist, nor will he be the last, who embraced the military paycheque enthusiastically to find new ways of destruction. Here he led the research which culminated in 1942 with the largest and most sophisticated rocket ever built. It was 46 feet high, weighed 27,000 pounds and reached a top speed of 3,500 mph. It also carried one ton of explosive to a target 200 miles away. This was the *Vergeltungswaffen-2,* 'weapon of retaliation'—the V-2. Technically it was a wonder, far ahead of anything ever built at that time, and it was to provide the basis for the American rocket and space programme. As a weapon it was not too efficient, but 5,000 of them were shot off and a number of people died in England.

Rocket and space travel societies had also been formed in America and Britain. The American Interplanetary Society was founded in 1930, but changed its name to the American Rocket Society in 1934 to attain the image of sober researchers rather than wild-eyed cranks. A number of its founders were identified with the emerging pulp field of science fiction. The first president, David Lasser, was managing editor of *Science Wonder Stories* and *Air Wonder Stories,* while the vice-president, G. Edward Pendray, wrote SF under the pseudonym of Gawain Edwards. Other founder-members were Laurence Manning, Fletcher Pratt and Nat Schachner, all well-known science fiction writers of the time.

The most important possible member stayed away, Goddard, still avoiding publicity, refused to join or provide the society with details of his experiments. The society did some experimenting, then concentrated on theory, and in 1963 it was one of the two organizations that merged to become the giant, ultra-respectable American Institute of Aeronautics and Astronautics.

In 1933 the British Interplanetary Society (BIS) was founded by P. E. Cleator. Like that of the American Rocket Society, the BIS's early membership included prominent members of the science fiction community. Among the editors of their early *Journals* and *Bulletins* was John Carnell, founder and editor of the magazine *New Worlds.* Others were Arthur C. Clarke, William F. Temple and Eric Frank Russell. Since the British Explosives Act made experimentation impossible the membership embarked on an ambitious theoretical project—nothing less than the design of a lunar spaceship. It was practical in every way, other than the decision to use solid fuel rather than liquid. The danger inherent in a fuel load of 2,000,000 pounds of gunpowder was expressed by the founder himself: 'I find it difficult to view with equanimity the prospect of careering moonwards atop a lighted powder magazine, however admirable the system of control.'

By the time real rockets were developed science fiction had left the moon, and even the planets, far behind. These were just milestones, stepping stones, on the way to the conquest of the galaxy.

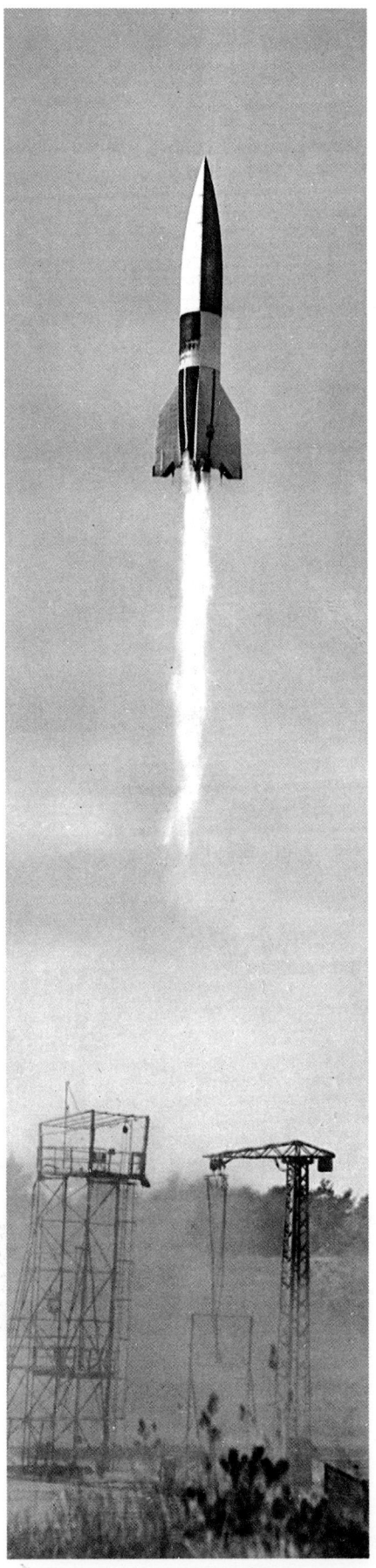

BEYOND THE SOLAR SYSTEM

Left *Frank R. Paul was the master of the science fiction spaceship, designing innumerable variations on the theme.*
Above *Howard Brown, at his best one of the most effective SF artists, launches one of 'Doc' Smith's famous space fleets.*

A copy of the first issue of *Amazing Stories* will cost you $200 today, but when it was published in 1926 you could have bought it for 25 cents. This was the first magazine to be devoted completely to science fiction. Although the slogan of the publisher, Hugo Gernsback, was 'Extravagant Fiction Today—Cold Fact Tomorrow!', there were few cold facts that ever emerged from its pulpish pages. But what this magazine did do was to liberate SF from the restraints of its conservative past. If Verne and Wells were to grace the pages of the first issues, they were soon to be replaced by wildly enthusiastic hacks with no knowledge of science and less of literature. The thing they did have was the power to entertain their juvenile audience; the stories were fun. The spirit of Tsiolkovsky and Goddard was behind them but they rushed into the future with a reckless abandon never known before.

Of all of the writers in the dawn of magazine science fiction, surely the most influential must be Edward Elmer Smith, Ph.D. Known as 'Doc' Smith to the generations of devoted readers who travelled with him to the remote corners of the galaxy, he launched all the space armadas that are still flashing through subspace. His first grand epic was *The Skylark of Space,* which had a hesitant beginning in 1915 with Doc's collaborator, Mrs Lee Hawkins Garby. She was to handle the 'love interest' which he felt himself incompetent to tackle. Mrs Garby did not last the course, however, and Doc finished the story himself in 1919. For the next eight years editor after editor rejected this classic-to-be (it's still in print) until the well-read manuscript crossed the desk of the newly founded *Amazing Stories.* It was eagerly bought by them, perhaps because the author agreed to accept only £75 for the undoubtedly dog-eared pages of his 90,000-word novel. This was one-sixth of their normal rates, which were already the lowest in the field, and was for all rights for ever. The novel was serialized in 1928 and became an instant hit with *Amazing*'s audience.

To the modern reader *The Skylark of Space* appears very badly written: the plot creaks, the characterization is more cardboard than wooden and the dialogue embarrassingly juvenile. None of this mattered in the dawn of the world, since most of the readers were juveniles themselves. They had had their fill by this time of backyard rocketships banging and clanking to Mars. With

Doc Smith they were out of the constraints of the solar system in seconds and enjoying the wonders of the entire galaxy. Travelling in vehicles that were something else again. Ships like the *Skylark*, which was:

... a spherical shell of hardened steel armorplate of great thickness, fully forty feet in diameter; though its true shape was not readily apparent from the inside, as it was divided into several compartments by horizontal floors or decks. In the exact center of the huge shell was a spherical network of enormous steel beams. Inside this sculpture could be seen a similar network which, mounted upon universal bearings, was free to revolve in any direction. This inner network was filled with machinery, surrounding a shining copper cylinder. From the outer network radiated six mighty supporting columns. These, branching as they neared the hull of the vessel, supported the power-plant and steering apparatus in the center and so strengthened the shell that the whole structure was nearly as strong as a solid steel ball.

This space-going ballbearing—or 'mighty vessel' as the author called it—was able to travel faster than the speed of light (goodbye, Dr Einstein) thanks to the hero's discovery of a method of liberating the atomic energy of copper atoms. Yet strong as the old *Skylark* was, she was to be immensely improved by the end of the book. The outer hull of *Skylark Two* is made of new metal, arenak, 'five hundred times as strong and hard as the strongest and hardest steel'. It is also transparent which alleviates the problem of trying to cut portholes in the stuff. When the hero becomes involved in a war on the planet Osnome that has been running for 6,000 years—everything is bigger and longer in Doc's work—he crashes his super-ship right through an enemy spacecraft. This is immortalized on the magazine's cover, the first of many exploding and crunching spaceships to come.

So happy were the readers with all this that Smith rushed back to his typewriter. By day he was the simple cereal chemist who spent a lifetime improving doughnut mixes, but by night he spanned galaxies. *Skylark Three* was serialized in 1930 and *Amazing* was indeed the title for a magazine that presented such wonders. There was interstellar warfare now, with ever-more-powerful energy weapons slicing the ether, while the hero, Richard Seaton, even had

Trying to help her, half kneeling over her, Dunark struggled, his green skin paling to a yellowish tinge at the touch of the bitter and unexpected cold.

Far left *Inside the* Skylark *showing the all-powerful drive and mighty supporting girders.*
Centre Skylark II *in action, blasting an enemy fleet to pieces by the most direct method.*
Left *It may look like glass, but this is the arenak hull of* Skylark II, *five hundred times stronger than steel and accordingly hard to scratch.*

a new spaceship. It was a 'ten-thousand-foot cruiser of the void ... one jointless, seamless structure of sparkling, transparent inoson'. How times have moved on! Within the two-mile length of *Skylark Three* is a little compartment where *Skylark Two* is tucked away in case it might be needed. Within two years of its construction the huge spaceship has become a midget.

In 1934 the third novel in the series, *Skylark of Valeron,* was published. A hint of panic can be detected in Smith's writing as he searches for even more superlative superlatives. All that he can say now about the newest ship is that she is 'fully twice the size of *Skylark Three* in every dimension'. He does not tell us if the previous ship is filed away in a corner. So in a few short years the spaceship industry has moved from 40-foot long ships to 4-mile long models. Even Detroit at its most exuberant could never equal this, for the author can write 'one mile' as easily as he writes 'one foot'. Once giantism sets in, the level of extravagance rises at an ever increasing pace.

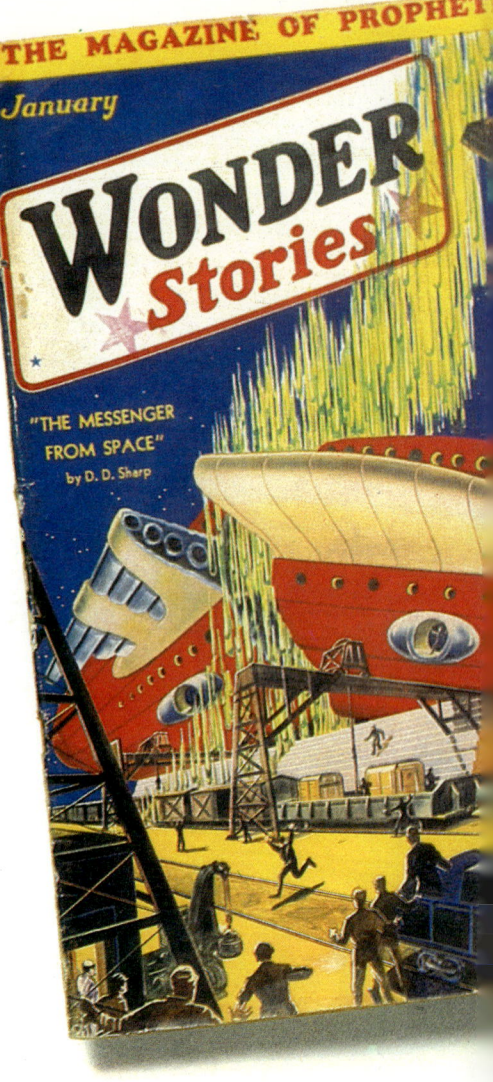

While Smith wrote about the future wonders, it was Frank R. Paul who made them live in his unforgettable artwork. Paul painted every cover for every science fiction magazine that Gernsback published in the twenties and thirties—as well as providing most of the interior art as well. All this as a sideline to his main job as a technical illustrator. Like Gernsback, from Luxembourg, Paul was also a European émigré. He was born in Austria and studied art in Vienna and Paris before moving to New York. There is more than one hint of Teutonic ancestry in his paintings; his nobly erect people could have stepped right off the set of *Metropolis.* Gernsback had been using Paul since 1914 to illustrate his electrical magazines, in which he published his first science fiction. Paul was the obvious man for this new medium of dramatized scientific wonders and, after a hesitant start, he reached full stride. Few of the stories lived up to the promise of the illustrations, for Paul could expand a single vague description into an entire alien technology—nuts, bolts, rivets, spinning wheels and fuming retorts. He created strange landscapes and cities that gave such shivers of delight to his thousands of happy readers that they still live strongly in memory. Though his spaceships were to be slavishly copied by generations of artists they were never equalled. They rarely floated peacefully by but were penetrated by meteorites, quartered by deadly rays or destroyed by spatial collision. It was all so good that it had to be true.

If Paul had any weaknesses it was with his people. His training in architecture and engineering produced some incredible technical works, but for many years he never seemed really at home with his figures. They many times stood aghast in one corner, he half-supporting she, as much impressed by the machinery as the reader was. All too often the hero wore engineer boots laced to the knee, topped by jodhpurs. But this is not to denigrate the great man; even God is allowed to nod occasionally. Paul was the artist who first showed the world what the creations of science fiction should look like. Therefore it was only fitting that at the first World Science Fiction Convention in 1939 Paul—not a writer or an editor— should be the Guest of Honour.

Doc Smith was not alone in creating giant spaceships in that dawn of magazine SF. In the very same month that *The Skylark of Space* began its serialization in *Amazing,* the magazine *Weird Tales*

BEYOND THE SOLAR SYSTEM   35

The great Paul at his best, not only a master of spaceship design but a creator of all kinds of space artifacts including an early space station (centre). The design is derived from an article by Captain Hermann Noordung of the VfR which was the first serious study of the possibility of building space stations.

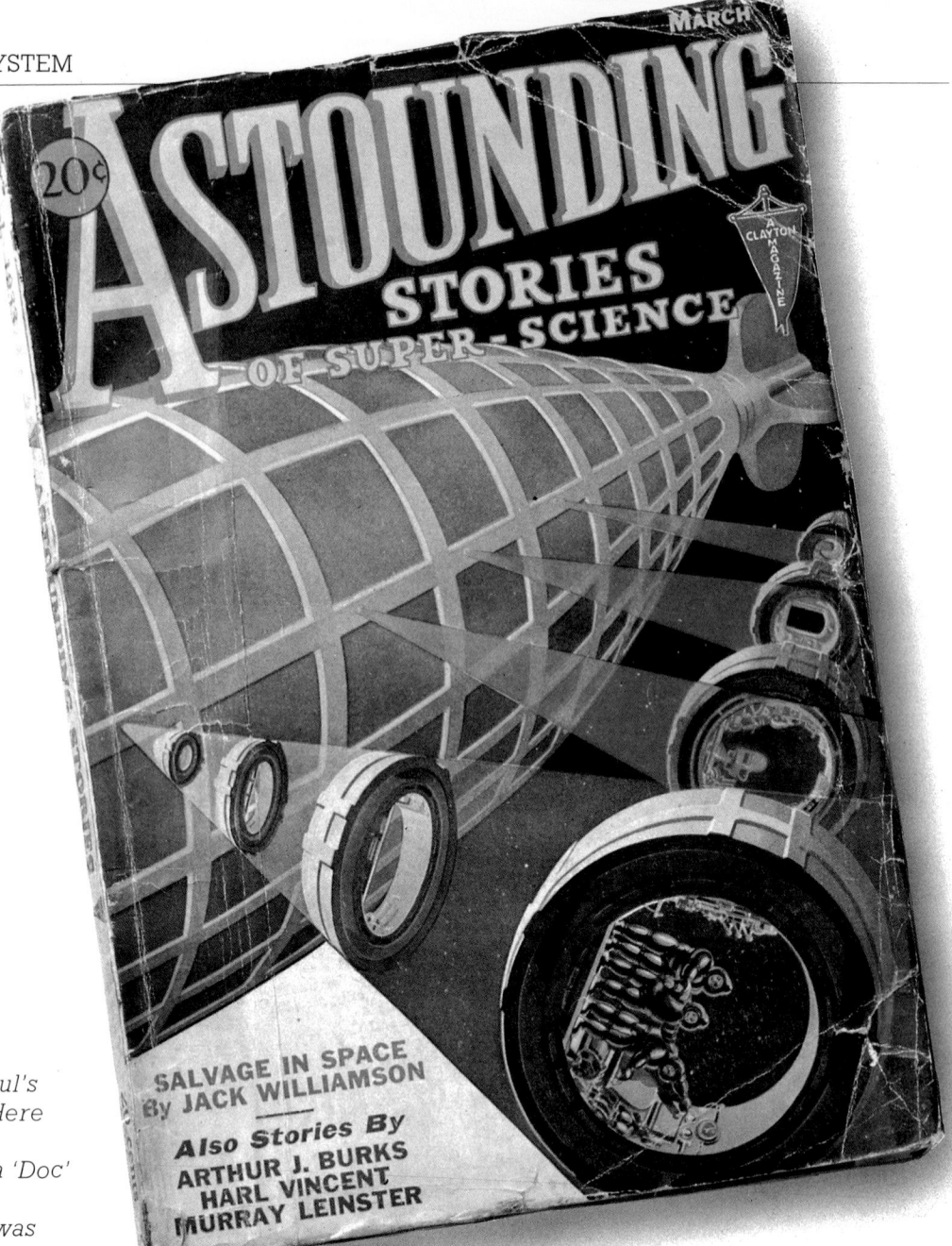

Wesso (Hans Wessolowski) invented his own brand of spaceship, different from Paul's but just as solidly fantastic. Here the lifeboats flee impending disaster. The scene is from a 'Doc' Smith novel which, through unfortunate circumstances, was not printed in the magazine.

published the first instalment of Edmond Hamilton's *Crashing Suns*. There was no false humility in the title—or in the writing—and the author soon became known as 'World-Saver' or 'World-Wrecker' Hamilton, depending upon the nature of the particular story. Hamilton's *Crashing Suns* gives us a good example of the slippery justifications that SF writers used to send their craft whistling about the galaxy well beyond the speed of light:

For the last three years Sarto Sen, one of our most brilliant young scientists, has been working on a great problem, the problem of using etheric vibrations as a propulsion force to speed matter through space. A chip floating in water can be propelled across the surface of the water by waves in it; then why should not matter likewise be propelled through space, through the ether, by means of waves or vibrations in that ether? Experimenting on this problem, Sarto Sen has been able to make small models which can be flashed through space . . . at a speed equal to the speed of light itself.

BEYOND THE SOLAR SYSTEM 37

Why not indeed! Full speed ahead for the ship and the story as well. The readers were more interested in non-stop action than a vindication of Einstein's theory of relativity. All the wondrous planets of the galaxy were waiting out there, so go we must. Though it did need a hard-working willing suspension of disbelief to get through some of the future wonders. For instance this spatially claustrophobic description of the navigational conditions around Earth:

I leaned out of the window, watching the dense masses of interplanetary shipping through which we were now threading our way. It seemed, indeed, that half the vessels in the solar system were assembled around and beneath us, so close-packed was the jam of traffic. There were mighty cargoships, their mile-long hulls filled with a thousand products of Earth . . . Sleek long passengerships flashed past us, their transparent upper-hulls giving us brief glimpses of the gay groups on their sunlit decks. Private pleasure boats were numerous, too, mostly affairs of gleaming white . . . Here and there through the confusion dashed the local police-boats of Earth, and I caught sight of one or two of the long black cruisers of the Inter-planetary Patrol . . . At last, though, after a slow, tortuous progress through the crowded upper levels, our craft had won through the jam of traffic.

Though we are well into the future, the prose is still in the biplane era. Take the first sentence, for example: it wouldn't be advisable to

Below left *The Midnight Mail takes off for Mars—a typically bold and stylized Elliott Dold design.*
Below *This cometary car by Wesso is unusually streamlined for 1932.*

Four automobile jacks were Ashembe's means of getting the inner chamber of his cometary car off the ground to place the plates for the base of the next shell beneath it—a system of fine arches carrying the weight of the inner projectile not more than two feet above the base of the other. . . .

BEYOND THE SOLAR SYSTEM

start leaning out of any *spaceship* window.

Doc Smith's greatest rival, however, was not Hamilton, but the young John W. Campbell, Jr. Campbell is remembered now for his 33-year stint as editor of *Astounding Science Fiction*—the title was changed to *Analog* during his era—but he first gained science fiction fame for his galaxy-traversing novels. He was working his way through Massachusetts Institute of Technology when he first started writing, and there is always a respect and knowledge of science lurking in the background of even his most exuberant novels. In 1931, in his *Islands of Space,* we see what a Campbellian spaceship looks like.

The great hull was two hundred feet long and thirty feet in diameter. The outer wall, one foot of solid lux metal, was separated from the inner, one-inch relux wall by a two-inch gap which would be evacuated in space. The two walls were joined in many places by small lux metal cross-braces. The windows consisted of spaces in the relux wall, allowing the occupants to see through the transparent lux hull.

From the outside, it was difficult to detect the exact outline of the ship, for the clear lux metal was practically invisible and the foot of it that surrounded the more visible part of the ship gave a curious optical illusion. The perfect reflecting ability of the relux made the inner hull difficult to see, too. It was more by absence than by presence that one detected it; it blotted out things behind it.

This optically disturbing craft is necessitated by reasons of plot, and the size of the galaxy, to go faster than light. One of the trio of heroes, Arcot, therefore quickly invents a faster-than-light drive,

Below *Frank R. Paul, father of the flying saucer? This 1931 space derelict bears more than a slight resemblance to the saucers of later sightings.*
Below right *Dold's alien spaceships (property of the Blue Men of Yrano) differ little from his human artifacts—plenty of portholes and scarcely any room for the engines.*

Hubert Rogers's spaceships convey a sense of power, size and strength and have a reality missing at times from the work of other artists. Here impressively solid spacecraft prepare for launch from their giant silos.

known affectionately in SF as FTL. Tinkering in some way with the nature of space, or discovering a 'hyperspace' where Einstein's equations do not apply, had become a needed device for getting around the theory of relativity. It should be noted that science fiction writers keep up to date on their rationalizations to get the plot moving to the stars, and today several of them postulate an FTL 'tachyon' drive. The tachyon theory proposes the existence of particles which actually cannot travel slower than the speed of light. Tachyons. Therefore if the switch can be pulled that changes all of a spaceship instantaneously to its tachyonic equivalent—away it would all go.

Despite all of the exuberant spaceship design, it is interesting to note that no SF writer ever correctly described the building and launching of the first spacecraft in reality. By the thirties the readers had had enough of backyard construction and firing-off of rockets and wanted their ships a mile long or more. Real spaceships were left behind in a cloud of rocket dust, to be painfully constructed years later by the engineers who had teethed on this wild fiction. There is a large hint that the magazine stories were more escapism than reality in the editorials of T. O'Conor Sloane, then the editor of *Amazing Stories*. At the same time that he was publishing Smith's and Campbell's interstellar epics he was stoutly denying the possibilities

40 BEYOND THE SOLAR SYSTEM

Right *The hero and heroine of* Things to Come *are lifted into position in the space shell, soon to be fired off to the stars.*
Below *Landscape showing the massive cannon and minute space shell hoisted by a crane.*

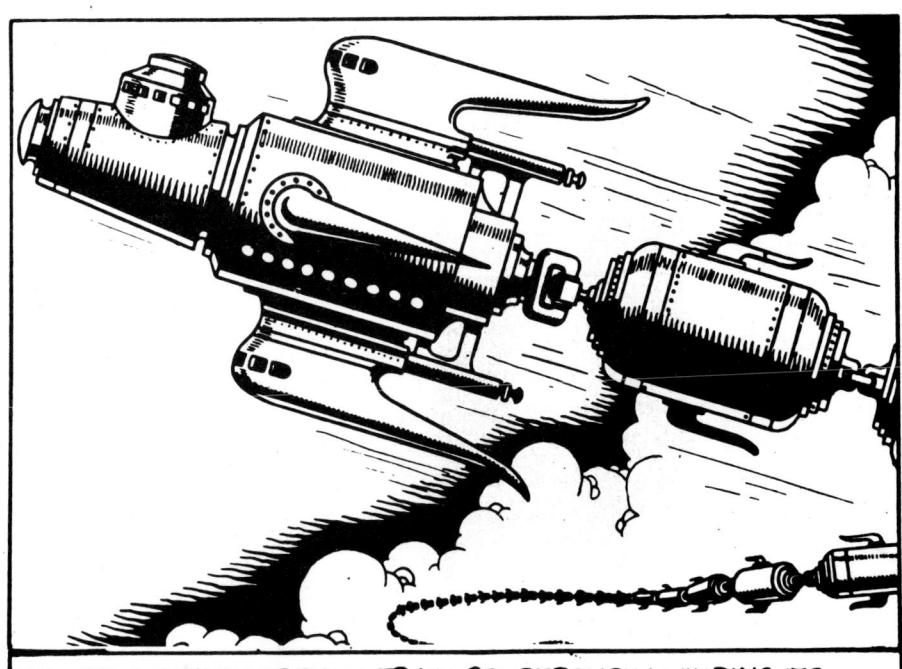

Dick Calkins, the artist who drew Buck Rogers, designed many an imaginative spaceship, including this rather lengthy space train.

of space travel in the opening pages of the magazine.

Nor was he alone in his lack of faith in real rockets, that they might actually blast in the near future. Olaf Stapledon, in his epic history of man's future, *Last and First Men,* thought so little of the chances that he placed the first spaceflight some 250 million years in the future. Since they have so much time to prepare for the flight, his people do it in the really grand manner. Inventing a suitable method of propulsion 'did not take many centuries' and the ship is built to circumnavigate the moon. Just like Apollo 8, only in slightly more grandiose fashion:

The first vessel to take to the ether was a cigar-shaped hull some three thousand feet long, and built of metals whose artificial atoms were incomparably more rigid than anything hitherto known. Batteries of 'rocket' apparatus at various points on the hull enabled the ship not only to travel forward, but to reverse, turn in any direction, or side-step. Windows of an artificial transparent element, scarcely less strong than the metal of the hull, enabled voyagers to look around them. Within there was ample accommodation for a hundred persons and their provisions for three years.

The concept of a 'side-stepping' spaceship is certainly a novel one that would be most interesting to see. More of these ships are built to transfer civilization to Venus after it is learned that the moon will soon crash into the Earth with the usual results.

H. G. Wells was still toying with the giant-cannon idea in the film *Things to Come* in 1936, though his barrels within cannon barrels at least looked a little more practical than Verne's. This masterful film was a landmark of the thirties SF cinema, standing very much alone. The only other space films of this decade were the *Buck Rogers* and *Flash Gordon* serials. They were made with low budgets, cheap actors and cardboard sets—but they thrilled a generation of

42 BEYOND THE SOLAR SYSTEM

Saturday morning filmgoers. Flash did so well that his further adventures were chronicled in *Flash Gordon's Trip to Mars* and *Flash Gordon Conquers the Universe*. Single-handed, too. Though these serials creak with age they are still grabbing the audiences on television. Who cannot be charmed by the rickety little rocket ships, their supporting wires showing, that bucket around with cigarette smoke rising up from their steam tubes?

The comic strips, which the films were based on, did much better in the line of rocket design. In his strip Buck has a choice of zipping about on the Jupiter to Mars flight in an aerial submarine that also does a nifty 50 miles per second under water, or going slower in a Trans-Spacianic Gyro-Blast freighter, or going even slower still in an Interplanetary Freight Train that tows 6,500 cars.

While Flash was born as a newspaper strip, Buck is a child of the SF pulps. His first adventures were chronicled in the August 1928 issue of *Amazing Stories*—a rare collectors' item because this was the issue that also had the first instalment of *The Skylark of Space*.

More and more new pulp magazines were published in the thirties and it was physically impossible for Frank R. Paul to illustrate them all. New artists appeared—Leo Morey, Hans ('Wesso') Wessolowski, Elliott Dold, Howard Brown and Hubert Rogers—and each of them had something new to add to spaceship design.

When Paul moved on to greener pastures it was Morey who

The planet Mongo, site of Flash Gordon's many adventures, hosts fleets of warring spaceships. Alex Raymond, a draughtsman of note, designed the rockets, which also featured in the Saturday-morning cinema serials.

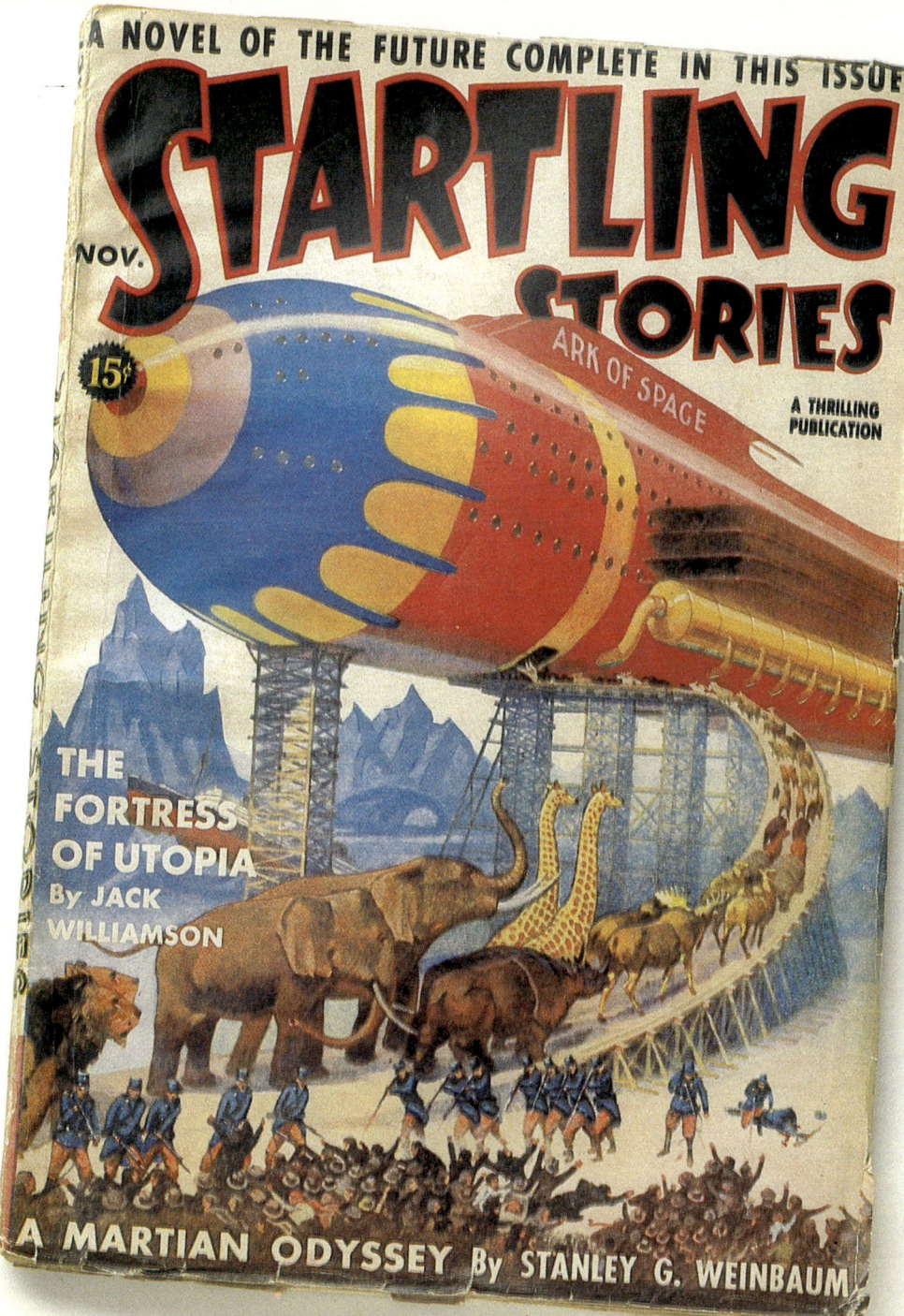

Howard Brown's justly famous space ark. Often the artist's imagination outstripped the author's. In Williamson's story the Ark is never completed.

had the unenviable task of filling his position on *Amazing*. Wesso was doing covers for the new *Astounding* while Dold was responsible for many of the fascinating black and whites inside. It was not until the late thirties that the Canadian artist Hubert Rogers began doing covers and interiors for *Astounding*. A highly trained painter, he captured the feel of machinery and spaceships, his spacecraft only equalled for size and authenticity by the superb black and white work of Schneeman.

The late thirties also saw the introduction of a totally new kind of spaceship: the Space Ark, or Generation Ship. Up until this time fantasy ruled the spacelanes and all of the 'interstellar drives' were more magic than science. Little by little there began to be some sober thinking among the writers as to how the great distances between the stars might be crossed. Without the FTL drive any voyage to the stars would take centuries rather than years. A possible solution to this problem is hibernation or suspended animation for the crew: into the freezer to wake up only upon arrival. An even more interesting possibility, since it makes for a better story line, is building a ship big enough to be a world unto itself. It would have to be enormous in order to hold the large population needed to maintain itself through many generations, as well as having a self-sufficient ecology. The classic early treatment of this theme is contained in two stories by Robert Heinlein, 'Universe' and 'Commonsense' (later amalgamated as *Orphans of the Sky*), first published in 1941. The idea itself, however, can be traced back to 1929 and J. D. Bernal's *The World, The Flesh and The Devil*. Bernal, a scientist of great distinction, was only 28 when he wrote this short monograph on mankind's possible future. He considers the varieties of possible futures for mankind in space, and eventually the possibility of journeying to the stars:

The difficulty involved in making this jump is probably as great as that of leaving the earth itself. Interstellar distances are so large that high velocities, approaching those of light, would be necessary; and though high velocities would be easy to attain—it being merely a matter of allowing acceleration to accumulate—they would expose the space vessels to very serious dangers particularly from dispersed meteoric bodies. A space vessel would, in fact, have to be a comet, ejecting from its anterior end a stream of gas which, meeting and vaporizing any matter in its path, would sweep it to the sides and behind in a luminous trail . . . Even with such velocities journeys would have to last for hundreds and thousands of years, and it would be necessary—if man remains as he is— for colonies of ancestors to start out who might expect the arrival of remote descendants. This would require a self-sacrifice and a perfection of educational method that we could hardly demand at the present. However, once acclimatized to space living, it is unlikely that man will stop until he has roamed over and colonized most of the sidereal universe or that even this will be the end.

In Heinlein's Generation Ship the 'perfection of educational method' is apparently not up to the expected level since there is a mutiny and a descent to a near barbaric level. His five-mile long ship is both universe and god to its inhabitants. The story must inevitably deal with the discovery of the true nature of the ship, as in other novels on this theme such as Aldiss's *Non-stop* (1958) and Harrison's *Captive Universe* (1970). It seems every SF author has a Generation Ship novel struggling to get out.

A more practical survey of the possibilities of travelling to the stars in this manner was written by Iain Nicolson in 1978. In *The Road to the Stars* he estimates that at least 10,000 people will be needed, disposes of the simple idea that because it is a ship it will have a crew and passengers, it will need instead some viable form of government, but concludes by saying, 'I cannot help feeling that the revolution in the human make-up required to make possible *successful* Ark missions may turn out to be greater than the technological hurdles which lie in the way of relativistic spaceflight!' It is hard to disagree, and difficult to imagine how 10,000 sane people could be persuaded to set out on such a mission. Of course 10,000 lunatics might be convinced to make the trip, but this story has yet to be written.

An unusual collaboration between two artists produced this graphic design of a space ark. Alex Schomburg (top) painted the exterior, while Frank R. Paul (right) delineated the multi-levelled interior, complete with shoemakers, sheep and hogs. Above Perhaps science fiction's first space colony. Morey's illustration to Jack Williamson's The Prince of Space *lacks an airlock—all the air is evacuated each time the port is opened.*

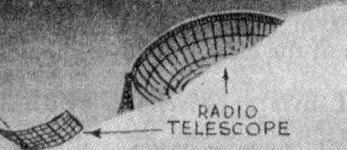

FICTION BECOMES FACT

We live in an age of momentous happenings. We have also been privileged to watch the most spectacular event in mankind's history, the leap into space; to see men walk on a world not our own. So we must not laugh too loud at Dr Richard Woolley who, as Astronomer Royal, said, in 1956, 'Space travel is utter bilge'. Poor Dr Woolley is not the only one who finds change hard to take.

A little over three decades ago the only viable rocket programme in the world was in shambles. When the pincer of the advancing Allied troops threatened Germany, the rocket installations at Peenemünde were evacuated and largely destroyed. The scientists there hoped that their rocket research would continue after the war—though obviously not in the fatherland. Since the prospects of a 40-hour week and a good salary looked better in the capitalist West than the socialist East their thoughts naturally turned in that direction. The ministries in Berlin were panicking and issuing contradictory orders. Von Braun and many of his associates selected the orders that would best suit their purposes. They disguised themselves as a mythical special project to divert the Gestapo's attention from what was obviously desertion in the face of the enemy, then headed south with their English dictionaries clutched tightly in their hands. They must have had uneasy consciences about all of the British corpses they had manufactured with their rockets for they did not surrender to Montgomery's army just 90 miles away on the banks of the Elbe, but instead continued southwards right through Germany to the Austrian Tyrol where they could surrender to the American army in May 1945. The world knows the happy results of this shrewd decision.

The war was nearly over in Europe, so the old rivalries were disinterred. The last thing the Americans wanted was the Russians getting ahead of them with rocket research, so when they captured an underground V-2 factory in Nordhausen they instantly realized its worth. They forgot to mention its existence to their Soviet allies, who were due to occupy this part of Germany soon, and instead sent a Colonel Holger Toftoy out with orders to pinch everything. The good

Vostok I, *the first rocket to carry a man safely into space, has 32 rocket chambers which must be fired simultaneously for liftoff.*

48 FICTION BECOMES FACT

A captured V-2 is prepared for firing by the Americans in 1946.

colonel was equal to the task and managed to move out 341 railway wagons of equipment, including a hundred rockets, with the last train rattling away just one day before the Russians arrived.

The Americans now had everything they needed to build the best rockets in the world; along with the engineers and the rockets themselves to experiment with. It is a monument to bureaucratic incompetence that even with this lead, and working steadily on rocket development, they were to lose every part of the early space race to the Soviets.

The first thing that went wrong for America was the rivalry between the U.S. Navy and the Army Air Force, both of which started rocket programmes after 1945. Instead of cooperating they worked independently and kept all their developments secret. Since the Army had captured von Braun in the first place they held on to him and shipped him and his V-2s off to White Sands, New Mexico, for military work and investigations of the upper atmosphere. By 1949 he had launched a V-2, with a Wac Corporal rocket on its nose as a second stage, that attained an altitude of 244 miles. This record remained unbroken for several years.

In the early 1950s there were a lot of things to think about, other than missiles. The United States Army, Air Force and Navy were all involved in rocket research, but on a small scale. These were all limited budget and limited aims programme. Since the large and unwieldy atomic warheads of the time were too heavy to load aboard a rocket for delivery, no research was being done into oversize military missiles that might be adapted as space launchers. However, when the first Soviet hydrogen bomb went off in 1953 Washington sat up and took notice. When this bomb was linked with reports of Soviet progress in developing ballistic missiles, very little pressure was needed from the military hawks to get their budget appropriations raised. The Army programme, which had moved from White Sands to Huntsville, Alabama, perfected the Jupiter rocket, which reached an altitude of 682 miles in 1956, with a maximum velocity of 13,000 mph. This launch could easily have put the first satellite into orbit. It did not. The final stage was a dummy—on the orders of the Defence Department.

This interesting turn of events was brought about by a decision which no doubt seemed wise at the time. As the United States thought they had the rocket game all to themselves, it was decided in 1955 to launch the first satellite ever during International Geophysical Year (July 1957 to December 1958). Not a bad idea, and the committee investigating the various rocket schemes examined those put forward by the Army, Navy and Air Force. The committee chairman, who was wiser about these things than the politicians on his committee, favoured von Braun's Army rocket. But he was overruled and the Navy's Project Vanguard was accepted. So the Army fired off successful duds and on 4 October 1957 the Soviets launched the first satellite, Sputnik 1. The entire American project became a great embarrassment because when the Navy pulled the switch on Vanguard, two months after Sputnik I, with great publicity, it simply sat there pouring out flame—then blew up.

Von Braun had never stopped work on his satellite proposal—and now he had the official green light. In January 1958

Above *James A. Van Allen* (centre) *triumphantly holds aloft a model of the rocket used in the experiment which discovered the radiation belt named after him. A beaming von Braun* (right) *lends more than moral support.*
Left *A Vanguard launch is prepared at Cape Canaveral.*

50 FICTION BECOMES FACT

Explorer I, America's first satellite, went into orbit. The space race was on. The Astronomer Royal had been wrong—space travel was not utter bilge.

Science fiction, which had invented the spaceship, now divided spaceflight into two categories. On one hand journeying to distant stars became commonplace. Spaceships were just part of the furniture, no more important than the car in which the bank robbers escape in an adventure novel. In the thirties the spaceships had been the real heroes of the stories, now they were just another bit of futuristic hardware. But in the other category SF rushed a few millennia backward in time to examine the near-future of the rockets being developed in the laboratories on Earth. In many ways these stories were fictional renderings of the ideas that von Braun and the

FICTION BECOMES FACT 51

others were bringing to fruition. Science fiction writers just stayed a few jumps ahead and showed where it all might be going.

Many times fact and fiction overlapped. Willy Ley, von Braun's old associate in the VfR, could not stomach Hitler and had gone to the United States when the dictator came to power. He was writing articles about rockets, for the SF and general magazines, and even produced two science fiction stories under a pen name. In 1951 he organized a symposium on spaceflight at the New York planetarium. This led to a series of articles in *Collier's* magazine, lavishly illustrated with the same rockets and space stations that used to adorn only the SF magazines. With public interest captured, books soon began to appear with titles like *Across the Space Frontier* and *Man on the Moon*.

Chesley Bonestell's detailed rendition of the ships for a moon expedition being assembled in orbit in Man on the Moon. *Inset The first of the ships touches down on the lunar surface.*

An early design for a three stage rocket from Across the Space Frontier, *painted by Rolf Klep. The vehicle is far more massive than any actually built: the manned compartment is the tiny section above the topmost fins; the rest is mostly fuel.*

Man on the Moon, with sober text by both Ley and von Braun, is interesting in that it gets the whole future history of rockets wrong. These specialist dreamers, who had worked with rockets all of their adult life, just did not realize that marked improvements could be made. They believed that chemically fuelled rockets could never develop enough thrust in a multi-stage rocket to reach the moon. They looked at the fuel-to-weight ratios that they were using at the time and never imagined how they would be improved. This book states that a vehicle built to make a non-stop flight to the moon would have to be 1,250 feet high and at take-off would weigh 800,000 tons. Or as much as ten ocean liners. Not a very practical suggestion. Arthur C. Clarke in 1951 in *The Exploration of Space* wrote more as an engineer than an SF author when he thought that such a vehicle would weigh 'millions of tons'. None of them predicted the more favourable mass ratios that would put men on the moon a decade later. Clarke even went on record as saying '... there is no possibility of fulfilling [the] requirements, for even the easiest of interplanetary return journeys, by the use of chemically-fuelled step rockets alone.'

With multi-stage rockets ruled out for the moon flight, the experts invented space stations as stepping stones. Von Braun designed one to be built at a height of 1,075 miles. It was to be a twin-spoked wheel, surely the cliché model for the SF space stations to come, and the prototype of the one in *2001: A Space Odyssey*. Von Braun's space station is built by three-stage rockets that ferry up all the materials from Earth. Once the thing is spinning along nicely three ships will take off from it in convoy to the moon, each of them weighing over 4,000 tons. The ships carry a total of 50 men. The whole operation is more like a naval battleship flotilla than the little space yacht that did make it to the moon in reality. Giantism still ruled the spacelanes.

After writing *The Exploration of Space,* Arthur C. Clarke put his later speculation in the form of realistic SF novels. These were sober extrapolations from known fact, and the author saw to it that his name on the title pages was followed by B.Sc., Fellow of the Royal Astronomical Society. *Islands in the Sky* in 1952 is a novel set in the future that had been described in his earlier non-fiction work. Clarke's space station is less streamlined and ordered than von Braun's Teutonic one:

My first impression of the Inner Station was one of complete chaos. Floating there in space about a mile away from our ship was a great open latticework of spidery girders, in the shape of a flat disc. Here and there on its surface were spherical buildings of varying sizes, connected to each other by tubes wide enough for men to travel through. In the centre of the disc was the largest sphere of all, dotted with tiny eyes of portholes and with dozens of radio antennae jutting from it in all directions.

Floating about the space station are spacecraft of various kinds including the unusual looking deep space vessels:

Others were the true ships of space—assembled here outside the atmosphere and designed to ferry loads from world to world without ever landing on any planet. They were weird, flimsy constructions, usually with a pressurized spherical chamber for the crew and passengers, and larger tanks for the fuel. There was no stream-lining, of

course: the cabins, fuel tanks and motors were simply linked together by thin struts. As I looked at these ships I couldn't help thinking of some very old magazines I'd once seen which showed our grandfathers' ideas of spaceships. They were all sleek, finned projectiles looking rather like bombs. The artists who drew those pictures would have been shocked by the reality: in fact, they would probably not have recognised these queer objects as spaceships at all . . .

In his 1951 novel *Prelude to Space,* Clarke deals with the problems involved in making the very first spaceflight. It is all quite complex with an atomic powered ramjet being launched in the middle of the Australian desert from a five-mile long track. On its back it carries the rocket powered vessel that makes it into space. The British engineers who have built this thing watch its successful take-off and comment smugly that 'The United States is much too small a country for astronautical research'. Indeed! A case of wishful thinking here, as the project has grown from the British Interplanetary Society of which Clarke was then President.

Even less realistically done was Robert Heinlein's 1947 novel *Rocketship Galileo.* Our old friend, the backyard American scientist, builds an atomic powered ship and zips off to the moon (with a singularly youthful crew as the book was aimed at teenagers). On the moon they discover a bunch of Nazi scientists who are going to hold the Earth to ransom from their stronghold and establish the thousand-year Reich at last. Clean living and American rugged individualism triumph as they always do in a Heinlein story.

Above *R. A. Smith's painstaking illustrations for* The Exploration of Space.
Left *A moon rocket is fuelled in preparation for its journey.*
Right *A space station is constructed and fuel ferried up to it.*

54 FICTION BECOMES FACT

FICTION BECOMES FACT

When George Pal wanted to make a moonflight film in 1948 he based it on this novel and summoned Heinlein to Hollywood. It was to be a serious film so the backyard scientists, and the Nazis and such, were junked and the very straightforward movie *Destination Moon* was made. It tells simply of the preparation for the first moon trip, the flight itself and the exploration after landing. Though the special effects look a little creaky now they were carefully done for the time, with Hermann Oberth brought over from Transylvania as technical adviser. The film was a success and reached a far larger audience than all of the books about the future in space put together.

One of the reasons was undoubtedly the set designs of Chesley Bonestell. This highly talented painter was over sixty when he embarked on a new career as an astronomical and space artist. He gained instant popularity with extraterrestrial landscapes for speculative books like *Man on the Moon.* His paintings are so convincing that they could pass as photographic records of space exploration. Just as Paul imposed his vision of the future on pulp science fiction of the twenties and thirties, so did Bonestell two decades later establish in the public mind just what spaceships and other worlds should look like. It is a tribute to the accuracy of his vision that the actual moon landing, as watched on television, seemed very familiar to many of the viewers.

There was a boom of SF films in the early 1950s, most of them pretty dreadful, but they featured an interesting array of spacecraft. In 1951 a select few citizens of Earth sought to escape the impending doom of *When Worlds Collide* by fleeing in a fair-sized spaceship. As in a *Prelude to Space,* this elegantly sleek and streamlined vehicle takes off from the end of a long ramp. The refugees land safely on their target planet and step out into an alien landscape that is an obviously painted backdrop. Equally alien, but much more impressive, was the strange craft of the title of the 1953 film, *It Came from Outer Space.*

Left *Chesley Bonestell's realistic set design for the film* Destination Moon.
Above *The spaceship in* When Worlds Collide *nearing completion on the launching ramp*.

56 FICTION BECOMES FACT

Left *The honeycombed spaceship in which* It Came From Outer Space *glows mysteriously in the Arizona night.*
Right *A 'close encounter' in* Earth vs. the Flying Saucers.
Below *The interior of the alien spaceship, complete with robot pilot and frightened girl, from* The Day the Earth Stood Still.

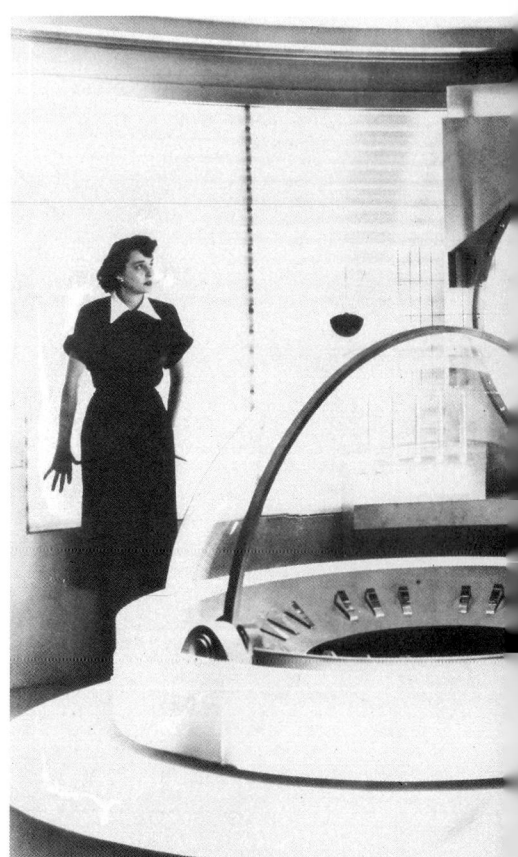

Most of the spaceships in the other films of the fifties, such as *This Island Earth* and *The Day the Earth Stood Still*, owe their design to a totally different source. Because these were spaceships bringing alien visitors to Earth—and by this time an alien spaceship meant only one thing to the general public. A flying saucer.

This craze for airborne crockery can be dated accurately to 1947 and an American businessman by the name of Kenneth Arnold. He was flying alone in his light plane one balmy June day, over the mountains of Washington, when he saw off to one side of him a string of shining objects. 'They flew close to the mountain tops in a diagonal chainlike line', he stated afterwards. 'I watched them for about three minutes. They were swerving in and out among the high mountain peaks. They were flat, like a pie-pan, and so shiny they reflected the sun like a mirror. I never saw anything so fast.' In a later version of his story he said that the flying objects looked more like saucers than pie-pans. This phrase was picked up by the press, eager to report unusual happenings, and entered the world's languages. No sooner had the first report been published than other people began seeing flying saucers too, and reports of unidentified flying objects have been coming in ever since. The saucer fad caught on and soon a new fringe cult had sprung up, to the joy of the publishers. In 1950 Donald Keyhoe, who had written a lot of rather pedestrian fiction for the lesser SF pulps, published a 'factual' account of saucer sightings. His book was bettered by George Adamski's which revealed that the author had not only seen the ships, but had watched them land and chatted with their alien pilots, then had actually been taken up for a spin as well. Adamski's book was 'documented' with photographs

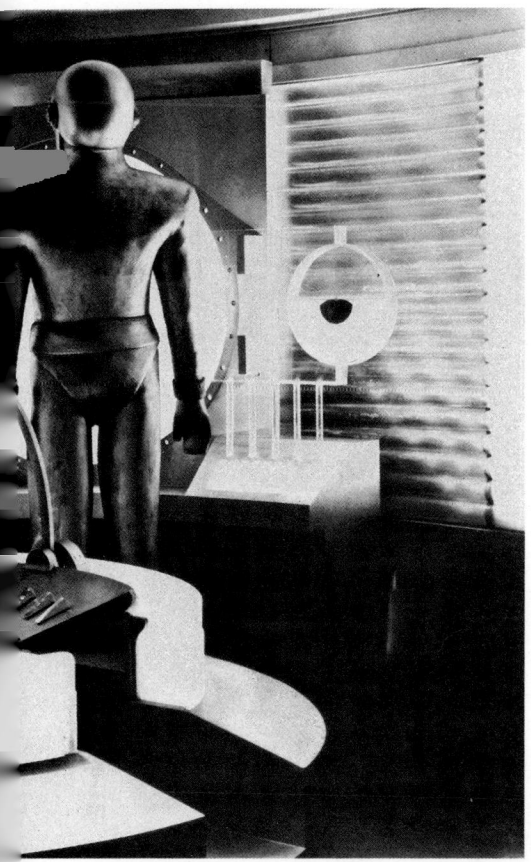

by the author who, despite his close encounter, had only been able to produce fuzzy blips of light. The few of his photographs that actually were in focus revealed an object that looked like a clobbered together collection of kitchen implements.

There seems to be little doubt, despite the 'sightings" magazines and books, that UFOs exist more in the minds of the beholders than in outer space. The saucer addicts bear a striking resemblance to the adherents of the Pacific cargo cults who also expected salvation from the skies. Science fiction writers and fans—far harder-headed and scientifically minded than one might suspect—universally dismissed saucers as one more crank cult. However, those responsible for magazine illustration evidently were not aware of the rejection of the theory and saucers appeared regularly on the covers after this.

The early 1950s also saw the death of the shabby, untrimmed-edge pulp magazines, which were succeeded by the neater, smaller and offset digest-size magazines. When the pulps died the exuberant and garish artwork died with them. The digests were more sober and serious about SF and science, and their covers tended to reflect their contents. While we all respect maturity, something went out of the world when the pulps passed on. No more the delights, on the covers of *Startling Stories* and *Thrilling Wonder Stories,* of Earle Bergey's fetishist fancies. Bergey's spacecraft were merely background props to the girl in the foreground, clad mostly in transparent plastic and a look of horror, being menaced by a Thing from outer space. For a while we also missed Alex Schomburg's highly imaginative continuations of rocket technology. He started

58 FICTION BECOMES FACT

Left *Alex Schomburg's flying saucers make a messy edge-on landing on the highway.*
Below *A fantastic spaceship from the ever-imaginative Frank Hampson.*

with adaptations of V-2 and von Braun's space station, then went on steadily into future developments. It is pleasing to note that he, alone among the pulp artists, is still active. His spacecraft still roar across the covers of the most popular SF magazine, *Analog*.

The digest magazines, such as *If*, *Imagination*, and *Fantasy and Science Fiction*, made attempts to have realistic covers. The last regularly featured Bonestell's alien landscapes, while artists like Ken Fagg and Malcolm Smith did their best in spaceship design. In Britain *New Worlds* followed the style with covers by Bob Clothier and Gerald Quinn.

To the youth of Britain in the 1950s, however, spaceships meant just one thing. *Anastasia*. Speak that word and eyes, once bright now rheumy, will glow again. For the generation of children who grew up then this was the craft of all time—not a missing member of the Romanov family but the trusty spaceship that ferried about Dan Dare and his chums. Every week in the *Eagle* they engaged the evil Mekon in battle and destroyed his slavering army

FICTION BECOMES FACT 59

Right *A panel from the* Eagle, *showing Dan Dare's famous ship, Anastasia, in action.*

60 FICTION BECOMES FACT

Above *A one-man spaceship with protective forcefield resists attack in this Malcolm Smith painting.* Above right *A flying city, protected by its spindizzy field, beautifully rendered by Solonevich.*

of Treens. Dan Dare was the most intelligent of British comics—and also the best drawn: Frank Hampson was an artist who took immense pains with his work and it was reflected in his popularity.

While 'realistic' science fiction was giving us a factual glimpse of the world of a day after tomorrow, the rest of SF was still blasting its way across the galaxy. Spacecraft of all sizes zipped from sun to sun on schedules as regular as ocean liners. And even bigger things were in store, the most grandiose of which was the spindizzy featured in James Blish's *Cities in Flight* novels. Blish's invention was properly titled the Dillon-Waggoner gravitron polarity generator, and more popularly called—for good reason—the spindizzy. This unusual device makes it possible to lift any object of any weight, protect it from all outside forces—and also move it along faster than the speed of light. The spindizzy is obviously an impossible invention, but Blish—one of the most knowledgeable and erudite of the SF writers—provides us with a theoretical justification, complete with equations, that instantly suspends our disbelief. Once conned we stay conned, for if you can lift anything why just stop at little spaceships? No reason. Blish's spindizzies lift entire cities and send them whistling off through the galaxy. The author postulates certain economic conditions and widely scattered human colonies that permit these flying cities to seek employment among the stars. Naturally one of these urban orbiters would not look much like spaceship, just a city above, ragged dirt below, with the entire thing

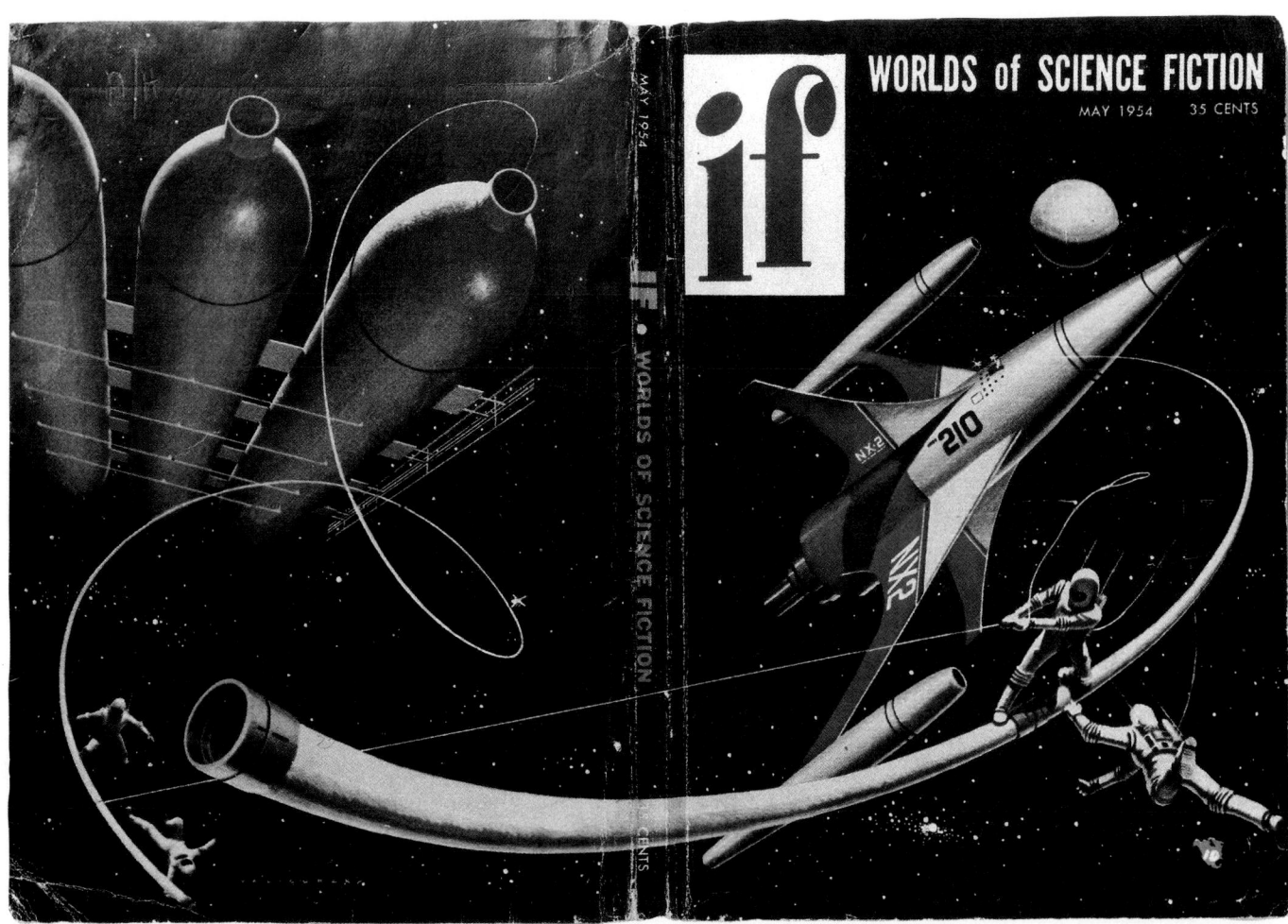

Above *Ed Valigursky illustrates the hazards of refuelling in space.*

enclosed in an energy bubble to hold in the air. But they would be impressive, particularly upon take-off, as described by Blish in *A Life for the Stars:*

After the first quake, however, Chris's alarm began to dwindle into amazement, for the movements of the ground were puny compared to what was going on before his eyes. The whole city seemed to be rocking heavily, like a ship in a storm. At one instant, the street ended in nothing but sky; at the next, Chris was staring at a wall of sheared earth, its rim looming cliff-like fifty feet or more above the new margin of the city . . .
  Now the city was level again, amidst an immense cloud of dust, through which Chris could see the landscape begin to move solemnly past him. The city had stopped rocking, and was now turning slowly. There was no longer even the slightest sensation of movement; the illusion that it was the valley that was revolving round the city was irresistible and more than a little dizzying . . .
  In a breath, the distant roadbed of the railway embankment was level with the end of the street; then the lip of the street was at the brow of the mountain; then with the treetops . . . and then there was nothing but blue sky, becoming rapidly darker.

While Blish's conception is a singularly fantastic one, he presents it in a matter-of-fact way that has the reader completely convinced. The visions of Cordwainer Smith, however, are unashamedly surreal and poles away from Blish's. Smith, the pen name of Dr Paul Linebarger,

Above *A sunjammer is readied for launching, portrayed by Alex Schomburg.*
Right *Don Davis's preconstruction of the NASA heliogyro. Halley's Comet in the background indicates an optimistic launch date of 1986.*

an American sinologist, political scientist and expert on psychological warfare, devised an elaborate and unique future history—featured in almost all his stories—in which mankind spreads across the galaxy in two sorts of highly unusual spacecraft—sailships and planoformers.

Sailships are the first to leave mother Earth. The first ones have sails 2,000 miles square, while the later ones become far larger. It is the pressure of light, solar energy, pushing against the sails that permits them to accelerate to a velocity to make an interstellar voyage possible. This is not an impossible idea, nor is it completely original. This idea was first proposed by Tsiolkovsky himself as early as 1920, as well as by another Russian scientist by the name of Tsander. It was also put forward—completely independently—by J. D. Bernal, who spoke of: 'A space vessel spreading its large, metallic wings, acres in extent, to the full, might be blown to the limit of Neptune's orbit. Then, to increase its speed, it would tack, close-hauled, down the gravitational field, spreading full sail again as it rushed past the sun.' This kind of vessel was given a name when, in 1964, Arthur C. Clarke and Poul Anderson coincidentally published stories, both of which were entitled 'Sunjammer'. Clarke's is the more dramatic version for he describes these gossamer-frail space yachts engaged in an Earth-Moon yacht race, the future equivalent of today's single-handed transatlantic races. Unlike terrestrial yachts there is a tremendous variety of design in space:

There they were, looking like strange silver flowers planted in the dark fields of space. The nearest, South America's *Santa Maria*, was only fifty miles away; it bore a resemblance to a boy's kite—but a kite more than a mile on its side. Further away, the University of Astrograd's *Lebedev* looked like a Maltese cross; the sails that formed the four arms could apparently be tilted for steering purposes. In contrast, the Federation of Australasia's *Woomera* was a simple parachute, four miles in circumference. General Spacecraft's *Arachne*, as its name suggested, looked like a spider-web—and had been built on the same principles, by robot shuttles spiralling out from a central point . . . And the Republic of Mars' *Sunbeam* was a flat ring, with a half-mile-wide hole in the centre, spinning slowly so that centrifugal force gave it stiffness. That was an old idea, but no one had ever made it work. Merton was fairly sure that the colonials would be in trouble when they started to turn.

There is a workable principle behind these craft, which was demonstrated by the unexpected orbit fluctuation experienced by the Echo I satellite. This was a big aluminized balloon in space, most sensitive to light pressure. Later NASA proposals included a solar sail and a species of 'heliogyro' which would be propelled by a dozen reflective and revolving blades each almost five miles long. The principle has already been applied in the Mariner 10 probe which flew by both Venus and Mercury. It used the pressure of solar radiation on its antennae and panels to conserve fuel and stabilize and steer the craft.

The sunjammer must surely be the most romantic spacecraft to sail through science fiction. Certainly it is the only one that truly can allow the vocabulary of sailing to be transferred to space. Even NASA's proposed ship of this kind has been christened the 'Yankee Clipper'. It is very unlikely, however, that such craft will really be

FICTION BECOMES FACT 63

useful in interstellar voyages since solar radiation falls off too quickly as you move away from its source.

Cordwainer Smith's other types of ship are far more exotic. Their pilots have the unusual ability of being able to 'attune' themselves to their ships, then, using their formidable psychic powers, they project themselves and their ships across the light years in an instant. His vocabulary echoes their strangeness, for these are 'jonasoidal' craft which travel to the stars by 'planoforming'. Incomprehensible as this is it still conforms to what Arthur C. Clarke calls his Third Law. This is that 'Any sufficiently advanced technology is indistinguishable from magic.' Smith's universe is one of the few convincing magical futures in science fiction. He has also written of what is possibly the largest spaceship ever invented. In 1959 in 'Golden the Ship Was—Oh! Oh! Oh!' there is a shimmering craft of the title that is 90 million miles long. Earth uses it to frighten off its potential enemies, who do not know that the thing is a gigantic fake made of thin foam and flown by a single man.

Undoubtedly the most singular spaceship of this period is found in Robert Sheckley's 1953 story 'Specialist'. This craft is a symbiotic construct of a number of different alien races. The living walls are made of linked-together, laminar creatures which have the ability to harden or soften their structure at will. Strange maritime

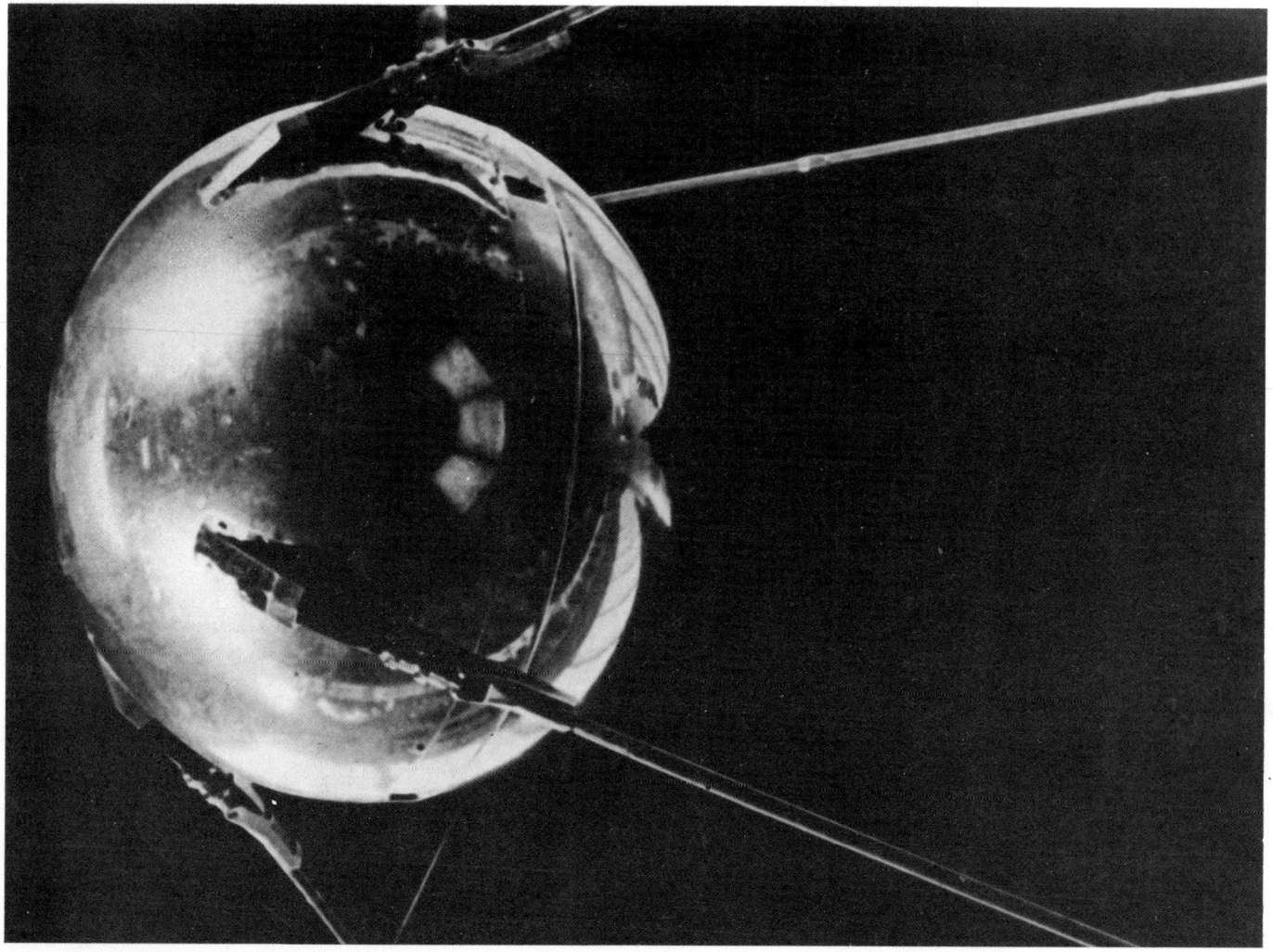

Sputnik One was little more than an instrument-packed ball with aerials attached.

creatures from a radioactive planet are the Engines, their stomachs forming atomic reactors. Other species make up the other functions of the ship, including the Pushers whose telepathic abilities boost the craft to velocities well beyond the speed of light. They are human beings. This living-machine has found pictorial representation in the works of the new artists of the seventies who specialize in anthropomorphic machinery.

While the fiction writers were far in the future, the Soviet space programme was very much in the present at this time. Their research into rocketry had not ceased with Tsiolkovsky's death; by that time government-sponsored research was well under way. As early as 1932 the young engineer Sergei P. Korolev had set up a programme directed by the chief of the army, Marshal Tukhachevsky. The first Russian liquid fuel rockets were tested in 1933.

That was the good news; the bad news was that the Stalinist purges were going on at this time. Tukhachevsky fell out of favour and was arrested in 1937 and later executed. Suspicion was automatically extended to all of his associates and many of them disappeared without a trace. Korolev had the good fortune, if one can call it that, of being associated with another suspect group, the aeronautic experts led by Tupolev who were accused of sabotage. This group was imprisoned, rather than killed like the Tukhachevsky people, so Korolev ended up in Siberia rather than in front of a firing squad. With the war on and designers needed for the various rocket missile systems, however, Korolev was later taken out of prison to join Tupolev near Moscow.

When World War II had ended Korolev was sent with Vladimir Glushko to Germany to study what bits of the V-2 programme had fallen into their hands. It was not much, for the Americans had made a pretty clean sweep. Yet there was enough to start a rocket testing programme in 1946 at the range in Kupustin Yar, very much like von Braun's work then under way at White Sands. Although he was still a suspect person, Korolev was put in charge of the tests, which continued for a number of years. They must have been successful because he was soon making proposals for long-range missiles. But real progress was not made until after Stalin's death in 1953. Korolev was free of suspicion at least, and even joined the Communist party for the first time at the age of 47. Khrushchev and the rest of the Politburo smiled favourably on Korolev and his plans for intercontinental ballistic missiles (ICBMs). In his memoirs Khrushchev remembers the rocket engineer in glowing terms: 'When he expounded his ideas, you could see passion burning in his eyes, and his reports were always models of clarity. He had unlimited energy, determination, and was a brilliant organizer.'

At this time the Soviet nuclear warheads were even larger and more unwieldy than the American ones, so a workable ICBM would have had to be bigger than anything the United States required. Korolev designed a hefty missile, the R-7, called Sapwood by the Western intelligence services. There were difficulties with R-7 from the start and many launch attempts failed. Yet by 1957 it had been successfully tested over a range of 4,000 miles.

When Korolev heard that the United States were going to launch a satellite during the International Geophysical Year, he suggested that the Soviets should do the same. His real interest had

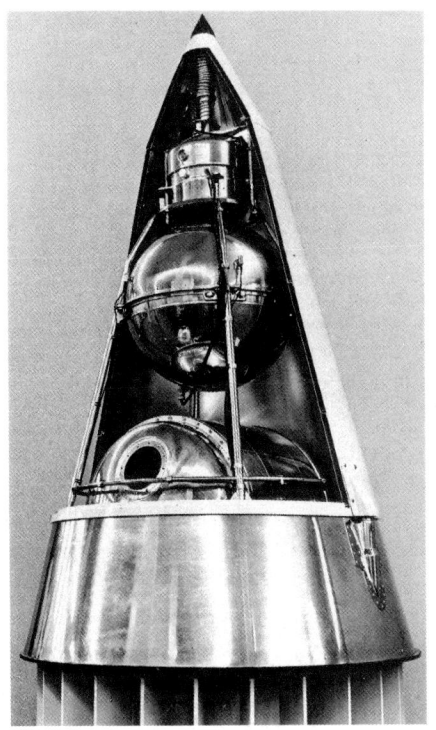

Sputnik Two nestles in the nose of the R-7 'Sapwood' rocket.

always been in space exploration and he knew that his R-7 had power enough to do the job. It was capable of lifting a payload of up to 4,400 pounds, which was far in advance of the American capacity at the time. There was some resistance to the idea but by the summer of 1957 it was approved. So on 4 October 1957 Sputnik I, weighing just over 194 pounds, was launched. The Americans were shocked by this, although they had plenty of advance warnings of the flight from the Soviets. They were even more rocked when Sputnik II took off a month later. It carried a then colossal payload of over 1,100 pounds—and also put the first living creature into orbit, Laika the dog. The satellites were in low orbits that decayed in a matter of months, but they were still orbits.

The giant R-7 was as useless as a missile as it was effective for space launches. At maximum range it could barely have made it across the Atlantic to hit the eastern coast of America. It was also slow and clumsy to prepare for launch, and difficult to guide. Korolev was more than pleased when future missile research was put in other hands and he was free to concentrate on the space programme. He was given a grand title that could have come from an old SF story—Chief Designer of Spaceships—and this was all that was known about him until after his death. Perhaps this was because Khrushchev loved to take all of the credit for the space successes himself. He had no desire to share the glory so saw to it that Korolev's role was played down. The ploy worked because his name is still little known in the West, although his importance to the Soviet space programme matches that of von Braun to the Americans.

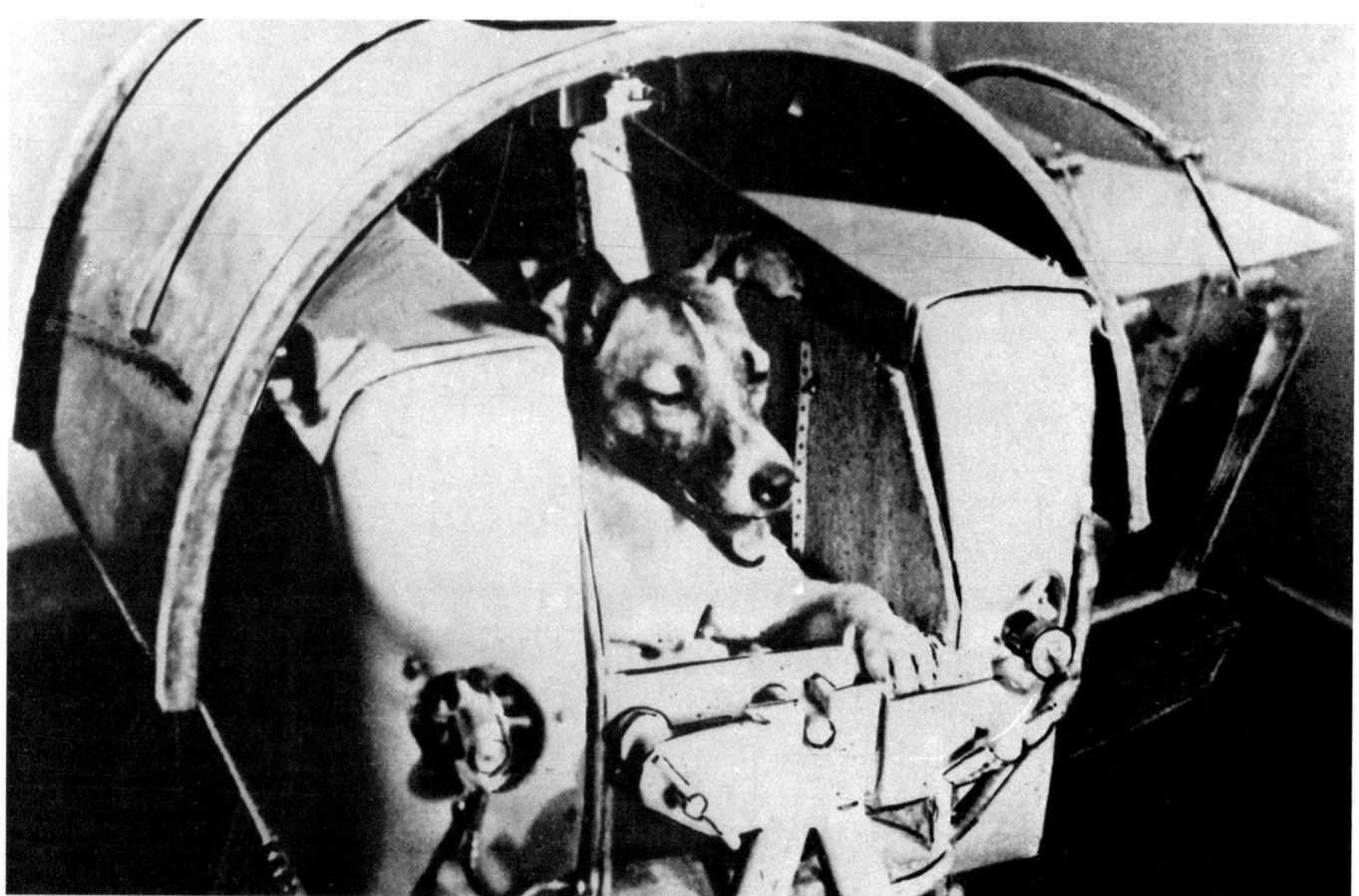

Spacedog Laika, the first living creature to go into orbit, inside the pressurized capsule of Sputnik Two. The dog lived for about a week before its air ran out.

FICTION BECOMES FACT 67

Above left *A Redstone rocket lifts Alan Shepard, America's first astronaut, from Cape Canaveral.* Above *Shepard's tiny Mercury capsule is lifted into position on the rocket's nose.*

The Americans were not happy at all about all this Russian space activity and saw it as a threat to their national pride. Funds were appropriated in astronomic quantities and the space race, if not the satellites, was launched. But the satellites did go up soon after and bigger and better carriers were being designed. At the same time the ICBMs were growing smaller so, like the Soviets, the American military and space programmes began to diverge. The civilian National Air and Space Administration—NASA—was created in 1958, and von Braun's Huntsville team joined this new body in 1960.

Project Mercury was a rush job in an attempt to give America some prestige in space. In 1958 development work was begun on the giant Saturn series of boosters which would eventually replace all of the other carriers. But that would be far in the future, so Project Mercury started at the same time using the already existing Redstone and Atlas rockets. These modified ICBMs could carry only a limited payload, so the Mercury spacecraft had to be very small: just nine feet high and six feet wide at the base. Getting a man into the thing was like slipping a banana back into its skin. By 1960 the test flights were ready to begin—without a human occupant—and the first two were failures. In January 1961 a Mercury capsule, after a series of mishaps, did carry a chimpanzee up in a ballistic flight, but not into orbit. The object was of course a manned flight, but once again the Soviets beat them to it. On 12 April 1961 Colonel Yuri Gagarin in Vostok I was launched from the space centre at Tyuratam. He completed one orbit of the Earth and landed safely after a flight that lasted 108 minutes. The first man had been sent into space.

Gagarin's flight had been carefully prepared by a series of trial Vostok launches. These were called Sputnik launches to mislead the Americans, just as the space centre at Tyuratam was said to be at

Left *Vostok spaceship and final rocket stage. The cosmonaut rides in the spherical capsule.*
Above *Two pioneers of Soviet space flight—Yuri Gagarin and Sergei Korolev*
Right *Inside the cramped Vostok capsule.*

FICTION BECOMES FACT 69

Baikonur. Vostoks were fired off in 1960 and 1961, but the first two were failures. Things improved after this and three more were launched, with dogs in capsules, and recovered safely. Gagarin's flight came shortly after the last of these tests. The Americans were well aware of what was happening in the Soviet Union—that was why they pushed their Redstone so fast—but it was to no avail. Vostok I took off and round two as well went to the Soviets.

The Vostok spacecraft that was put into orbit was far bigger than the American Mercury, over 24 feet long. This included the cylindrical final stage rocket which carried observation instruments, as well as a seven-foot diameter capsule for the pilot. For re-entry Gagarin separated the capsule leaving the rest in orbit. He actually landed inside this space-going sphere, which could not have been too comfortable since it made the last stage of the descent dangling from a parachute—coming down on hard ground. He never reported what it felt like but it is worth noting that all of the later cosmonauts ejected at 20,000 feet and landed by parachute.

Three weeks later the Americans got their man into space, if not into orbit. Alan Shepard rode up in a Mercury capsule named Freedom 7, perched on the nose of a Redstone rocket. His ballistic flight lasted just 15 minutes and hit a maximum height of over 117 miles. Once again the Americans had come a poor second in what they were calling the space race. They did not like losing and within the next decade they would overtake and pass the Soviets in manned spaceflight.

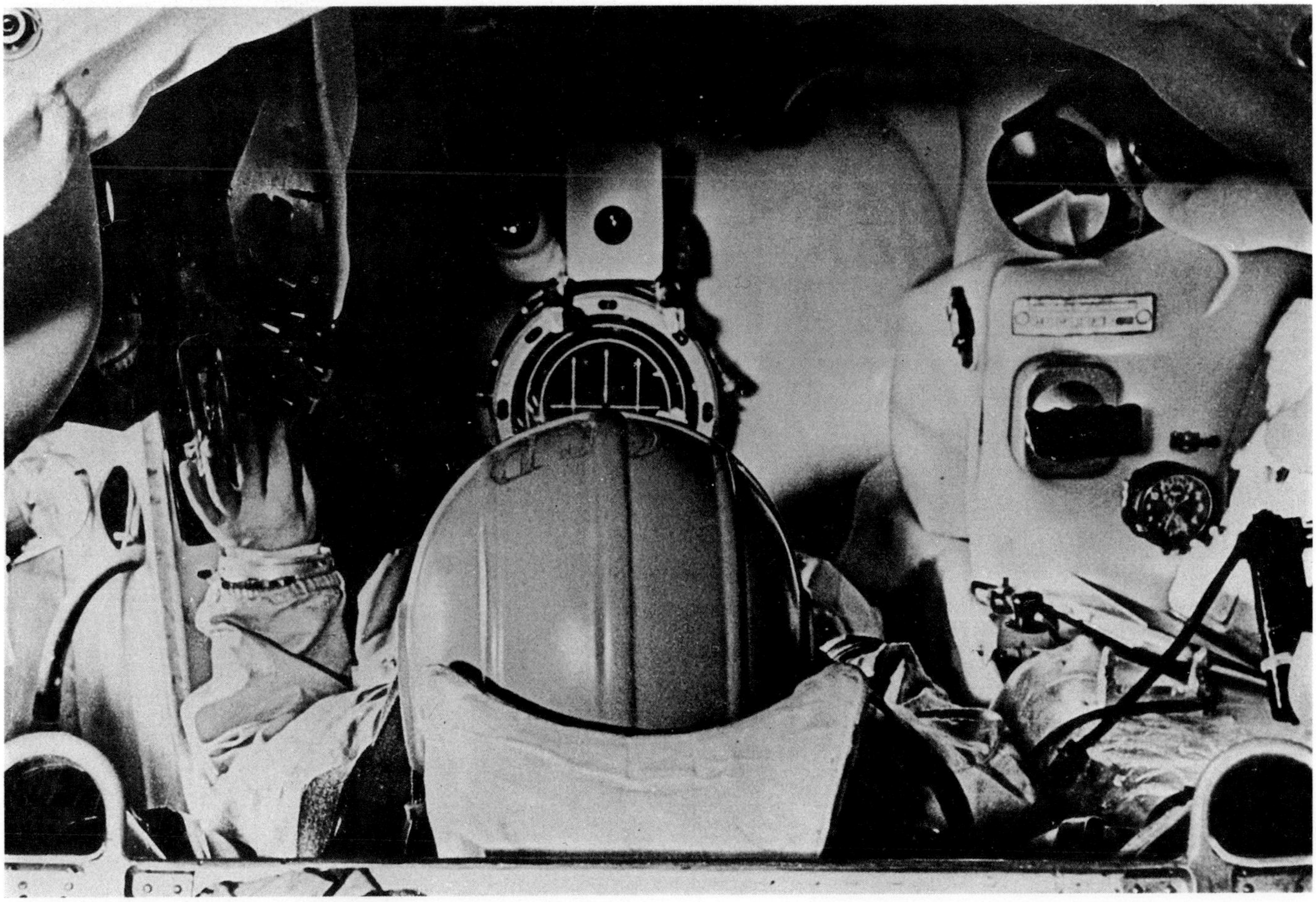

THE SPACE AGE

The race to the moon was on. Both the Americans and the Soviets had put a man into space. The next big target was the moon, though the Soviets later claimed that they never had any intention of manned lunar flights. Perhaps true, perhaps not—but in either case they were ahead of the Mercury programme with their Vostoks and bigger lifting vehicles. While America did not look too good in space, they were already pushing a massive research and development programme aimed at the moon. The Apollo spaceship was being designed, as well as the massive Saturn V launch vehicle that would carry it into space. At the same time the Soviet programme was encountering difficulties, partly caused by Khrushchev using the space race to boost his own prestige, as well as that of the Soviet Union.

1962 saw another first when Vostoks 3 and 4 made the first simultaneous flights, with cosmonauts Nikolayev and Popovich staying in orbit for nearly three days. They did it again in 1963 when Vostoks 5 and 6 also orbited in conjunction. This was a first in a different way. One of the cosmonauts, Bykovsky, is just a name in the books. But the other, Valentina Tereshkova, will be remembered as the first woman in space. She did experience some difficulties during the flight, and rumour has it that she was really a last-minute replacement for a highly trained woman who fell ill. And still other records were broken. Nikolayev stayed in orbit for over 94 hours, and that record was soon surpassed by Bykovsky, who managed 81 orbits for a record 119 hours and 6 minutes. The Americans were not quite catching up, but were nevertheless plodding on relentlessly. On 20 February 1962 a Mercury finally got into orbit with John Glenn, making three orbits, and launching him as well into a political career. After three more Mercury missions astronaut Gordon Cooper raised the record to 22 orbits.

The Soviets were still out for records and on 12 October 1964 they set another one with Voskhod 1, a new spaceship that could carry a crew of three. There was plenty of publicity for the cosmonauts, Komarov, Yegorov and Feoktistov, but very few details about Voskhod itself. It was not until photographs were released, much later, that its secret was known. This lack of information was not noticed at the time for just as Voskhod 1 was going up Khrushchev was going down for the last time. Even as he was talking to the

Earthrise—lunar module Eagle *approaching the command module of Apollo 11.*

cosmonauts the decision was being made to remove him from office. The space programme went on without him and there was another first in March 1965 when Alexei Leonov made the first space walk from Voskhod 2.

The Americans were slowly catching up, however. The first Gemini mission in 1965 carried astronauts Grissom and Young. This craft was far superior to the space-going bathtub that was the Mercury. There was even an element of thrust control over Gemini, supported by a very important first: the computer on board. The pilots were no longer just passengers, as incapable of affecting their flight as the first chimp that rode one of the capsules. They had control. It was years later before it was discovered that while the Americans were advancing the Soviets were just standing still. Photographs eventually showed that the marvellous 'new' Voskhod was really just the old Vostok with extra couches jammed in. It was Khrushchev himself who had ordered this, in order to stay ahead of the Americans. In a sense it was a step backward, three men crammed into a seven-foot space for political not scientific value. They could not wear the bulky spacesuits, so the entire cabin had to be pressurized. More important, all the safety measures had to be removed because of the extra weight. There were neither ejector seats nor an escape tower to carry them clear of the rocket in case of launch failure. Had anything gone wrong with either of the Voskhod launches the cosmonauts would never have stood a chance. Korolev, the designer, was most unhappy, to say the least, that all the safety measures were compromised for propaganda effect.

Though it can only be seen with hindsight, the American space programme was now in the lead. All of the Soviet space shots

Far left *Alexei Leonov enjoying man's first walk in space.*
Left *Project Gemini is launched! This unmanned test took place in April 1961.*
Above *Gemini 6 as seen from Gemini 7 during the first successful space docking manoeuvre.*

had been made with variants of Korolev's original R-7 booster. The final modifications could and did put seven and a half tons of payload into orbit, which was a considerable advance over the first American launch vehicles—but the Americans were not standing still either. The Saturn 1B and the Titan rockets could lift more than the ageing R-7s, while the enormous Saturn V outstripped all of them. The Soviets were slowly developing a large booster of their own, the Proton, which would be able to lift 22 tons when it was launched in 1965. The Saturn V was still larger.

It was too late for Korolev. He never lived to see the final development of his programme free of Khrushchev and his publicity stunts. He died at 59 of heart failure following routine surgery. Evidently there was no one to fill his shoes for with him died the dynamic drive of the Soviet space programme.

Quite the opposite was happening in America, however. The Gemini programme was a great advance on the tiny Mercury capsules, with a craft 17 feet long and 9 across. Ed White finally made the first American space walk in Gemini 4, and Borman and Lovell set a new endurance record in Gemini 7. They stayed up for 206 orbits and were in space for over 302 hours. There was trouble with the sixth flight—but this was also the first successful docking in space. The programme went ahead at a remarkable pace all through 1965 and '66 with a total of ten flights, nine different docking manoeuvres and space walks by three astronauts. Buzz Aldrin's walks lasted all of five and a half hours. The astronauts gained experience, hardware

74 THE SPACE AGE

was designed and improved, and all the techniques of launching a vehicle, getting it into orbit and then recovering it safely were perfected. They were ready for the next step: Apollo.

An unmanned Apollo Command Module was placed in orbit in 1964, riding up on one of the giant Saturn V boosters. Success seemed just a step away when a sudden and tragic fire in January 1967 incinerated three astronauts. A year and a half was lost while the cause of the fire was determined and Apollo redesigned and reconstructed so that this could never happen again. Then, on 11 October 1968, Apollo 7 went into orbit with astronauts Schirra, Eisele and Cunningham aboard. It stayed in orbit for nearly 11 days.

The Apollo spacecraft was as different from the first Mercury capsule as an articulated lorry is from a sports car. A single man was jammed, unmoving, into Mercury. But three men travelled in Apollo, which was over 75 feet long and weighed 50 tons. It was made up of five parts, only three of which went into orbit. The escape tower, which would have lifted the astronauts to safety in case of trouble on

Below *There is no up or down in space, although the lunar module here appears to have flipped over during an Apollo 9 test manoeuvre.*
Right *Apollo 10's command module over the far side of the Moon, photographed from the lunar module.*
Below right *Saturn V and the Apollo spaceship.*

THE SPACE AGE 75

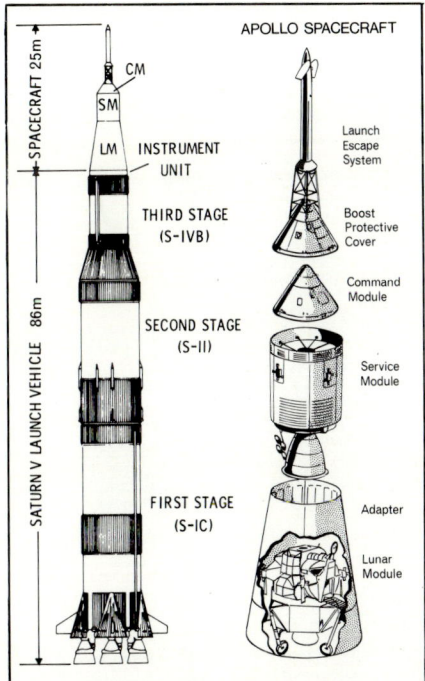

liftoff, was blasted free first. Once the craft was out of the atmosphere the aerodynamic shield was jettisoned as well. What remained were three completely separate, yet coupled together, spacecraft. The smallest, less than 6 tons and 11 feet long, was the command module in which the crew spent most of their time. The service module contained the propulsion system and was the largest, over half the total weight. The part of the payload that was really different this time was the lunar module. If all worked well this was the unit that would land on the moon. Once down it would be split in half for take-off, leaving behind the landing gear and exhausted rocket system while the ascent stage lifted back up to rejoin the command module. When Apollo 7 was launched safely into orbit it meant that the next objective had to be the moon.

While the first steps of the American space programme had been cautious, a step-by-step approach, the later developments came with a sudden rush. Within a few months of this first manned Apollo launch another Apollo went into orbit around the moon.

Borman, Lovell and Anders were the first men ever to see the back of Earth's satellite. But why the rush? It was politics again, the same kind of thing that had got the Soviet space programme into trouble. The war in Vietnam was at its absolute worst in 1968 and the space programme looked like an expensive luxury to a lot of people. Not only that, but there was talk that the Soviets were also planning a flight at this time to put one of their cosmonauts into lunar orbit. America had to beat them; another Soviet 'first' at this time might put paid to all of the American involvement in space. The leisurely development schedule was scrapped and the first lunar circumnavigation was made. The Soviets were beaten at a crucial time.

They were working hard on the new Soyuz programme, that much was known. Despite all of the Soviet disclaimers to the contrary it was also well known that they had the desire and the capacity to get a manned flight into lunar orbit and return. They did not plan to land—they had no equivalent of the American lunar module—but with the development of the Proton launcher they had all they needed to get there. This giant would first put a rocket stage up into orbit. Then the reliable old R-7 would send up a Soyuz spaceship. These two would dock in space. Then the moon.

Soyuz was 33 feet long and divided into three sections like Apollo. The service module would do all the pushing. The cosmonauts would work and sleep in the orbital module during the trip. Upon return they would move to the re-entry module, jettison the other two units, and drop back to Earth. The first Soyuz flight in 1967 was less than successful. There were problems with the new craft during the mission, and it ended in disaster when Vladimir Komarov was killed after re-entry when his parachute became tangled. Khrushchev's emphasis on the ego-building Voskhod project had already delayed Soyuz for a year, and these new problems lost another eighteen months. But the preparations still went ahead and the Soviets appeared to have the capacity to send a manned craft

Soyuz spacecraft on the launchpad.

around the moon by the end of 1968. This was the year of celebrations for the fiftieth anniversary of the Russian Revolution. A moon shot would be a fitting climax to the festivities. No wonder the Americans were worried—even more so after two Zond (unmanned Soyuz) missions actually did circle the moon and return. The next step was obvious. There was a good chance that the pilot would be Yuri Gagarin, but he was killed in a plane crash early in the year. But the pilot didn't matter; defeat did.

The lights burned all night at NASA and the Apollo 8 flight was changed. It was only supposed to have been an Earth orbit, like Apollo 7. Instead it was headed for the moon. So were the Soviets, who had the booster and craft on the launch-pad in December 1968 with an astronaut ready to go. Their launch window, the time they would have to blast off to get into lunar orbit, was earlier than the American window on the other side of the world. They would get there first. A heavy snowfall was said to have blocked their efforts, but we may never learn the truth for the launch did not take place. Apollo 8 did the successful flight instead and manned lunar flights were dropped from the Soviet plans; too risky and expensive, they said. Sour grapes, said others. In either case the road ahead was now clear for the Americans to make the biggest jump of all. To leave Earth, to land on another world, to walk the surface of a world other than the Earth for the first time.

There were two more preparatory missions to prepare and train for the big flight. Then on 16 July 1969 Apollo 11 took off. The

Rendezvous in space as Soyuz 19 nears Apollo during the historic joint flight. No faces can be seen behind the clearly visible portholes.

78 THE SPACE AGE

flight was smooth, on schedule all the way. En route to the moon the lunar lander was detached and its position reversed on the nose of the Apollo craft so it proceeded with its legs extended. It was the first deep-space craft, ugly and angular and destined never to enter a planet's atmosphere. On 20 July, when they were in orbit around the moon, Neil Armstrong and Buzz Aldrin entered the lunar module and disconnected from the command module. When they fired their retrorockets this vehicle, named the *Eagle,* dropped in a controlled descent to the moon's surface. The landing was dramatic enough, for despite all of the advance planning for a smooth landing site they found themselves heading for a rocky area. Armstrong cut off the automatic landing controls and flew the *Eagle* himself, aiming for a smoother region. He set her down with about 20 seconds worth of fuel in the tanks. No science fiction writer could have cut it finer.

In a dramatic moment, watched by millions on television, Armstrong emerged from the craft and stepped down on to the

Neil Armstrong and Buzz Aldrin in the Apollo 11 command module.

Above *An Apollo spacecraft is eased into position on top of a Saturn V rocket. The colossal Vehicle Assembly Building at Cape Canaveral—the world's largest building—looks like an extravagant science fiction movie set.*
Above right *The first men on the Moon set up a scientific experiment.*

moon's surface. The Americans had done it and who could blame them for opening out a flag—stiffened by wire in the lack of any atmosphere to flap it. Many would agree with Olaf Stapledon who predicted twenty years before this date, in 1948, something of this sort in "Interplanetary Man": 'Alas! Must the first flag to be planted beyond the Earth's confines be the Stars and Stripes, and not the banner of a united humanity?' Any other country would have done the same. But it should not be forgotten that the United States made no claim to any part of the moon after the flag-raising—unlike every other first arrival in history—and Neil Armstrong was quite sincere when he spoke at the time: 'That's one small step for a man; one giant leap for mankind.'

80 THE SPACE AGE

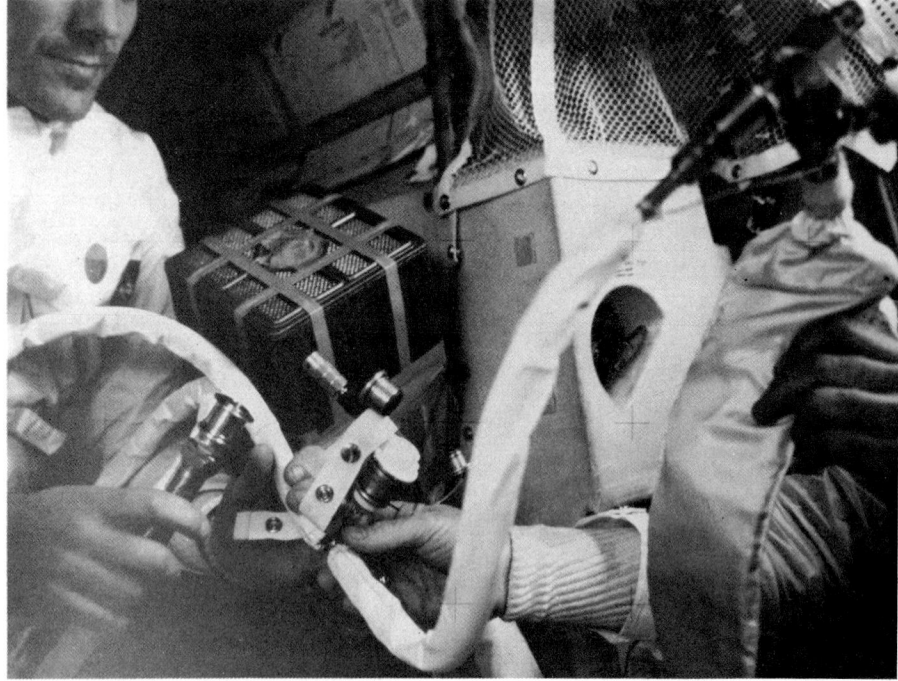

Right *Inside the Apollo 13 lunar module, the astronauts rig up emergency air-conditioning using air purification canisters from the stricken command module.*
Below *Apollo 13 sets off on its dramatic journey.*

A giant leap it indeed was. For two and a half hours the two astronauts collected a truly priceless collection of lunar samples, founding the new science of lunar geology on the spot. They then returned to the lunar module, disconnected the landing gear and exhausted landing engine, and took off. (Later Apollo missions left TV cameras behind, making it possible to witness the incredible scene as the disc-like and totally impossible looking craft was launched.) It was a moment to remember—for the world as well as for them. They reunited with the command module and returned to Earth to a fitting heroes' welcome.

Science fiction had predicted this landing often enough, but had never got it right. It was always a private thing, noble spaceman stepping to the lunar dust in airless solitude. No one had predicted that the entire world would be watching.

It is a testament to the immense amount of work, planning, testing and design, that five more lunar landings went off without a hitch. They may have looked commonplace to the unthinking observer but they are a lasting memorial to technological man, thinking man. On his own a man can leap five, six feet into the air. That's it. By the application of intelligence we have reached the moon and walked upon it. The event itself contained all the drama and excitement found in the type of fiction that sponsored it. The idea must come first, then the actuality. Science fiction invented space travel. And inspired by the idea men have walked on the moon.

Things were not quite so lucky for Apollo 13. The superstitious had a field day with this one for its launch time at the Manned Space Center in Houston was 13.13 hours and disaster struck on the 13th of April. (Of course although it was 1.13 PM in Houston, it was a completely different time in Florida when the craft took off. But it never pays to argue with superstition.) When the craft was 205,000 miles from Earth there was a sudden explosion. A faulty oxygen tank on the service module had exploded, tearing the module open.

THE SPACE AGE 81

There were only 15 minutes of power left in the command module, a disaster from an SF story that was actually happening. It was believed that there was only enough oxygen to last the astronauts for just half of the remaining voyage.

This was drama indeed, and the men in the ship and the scientists back on Earth were up to the challenge. They were committed to the moon orbit and could not leave it. They must circle the moon and use its gravity to get them safely back in the correct orbit to Earth. The ship would make it—but would the astronauts survive?

It is a tribute both to the astronauts in space and to the engineers on the ground that disaster was averted. Supplies were stretched—at very considerable discomfort—and the crew moved to the lunar module as an emergency craft, firing its engines to make course corrections. Emergency air-conditioning had to be put together quickly since the lunar module's machinery was not designed for this heavy a load and the carbon dioxide level had to be kept down. The craft circled the moon and started back towards Earth. After three and a half anxious days they returned to a safe landing. This rescue operation had occupied thousands of technicians and endless hours of computer time—far different from the

After the Apollo 13 crew had blasted free of the service module they could see the extent of the explosion.

classic science fiction disaster in space which the captain puts right with his trusty tool-kit.

Though the Soviets had abandoned the idea of putting a man on the moon, they were still deeply involved in a lunar programme. More Zond craft made flights around the moon and back, and even before Apollo 11 began its journey they had put Luna 15 into orbit around the moon. A soft landing was attempted but there was a malfunction and it crashed instead. Yet there were successes to come. On 12 September 1970 Luna 16 made a soft landing, scooped in some soil samples, then took off safely and returned to Earth. On 10 November Luna 17 also soft-landed but with far more spectacular result. A door opened in its side and a ramp slid out. Down this rattled a most unusual machine, Lunokhod-1, that looked as though it had driven right out of a Frank R. Paul painting. It had eight little wire mesh wheels below and television cameras, aerials and a laser reflector sticking out above. As soon as it was clear of the ship, Lunokhod stopped and flipped its lid, the entire top opening on a hinge to reveal the solar cells inside. Strange as the thing looked it was a most practical machine, and spent eleven successful months rolling about the lunar surface and taking tests with an X-ray spectrometer and a penetrometer mounted between its wheels.

Above *Eugene Cernan prepares for a ride in the Lunar Rover—or moon buggy—which later Apollo missions took with them.*
Above right *Artist's impression of Luna 17 touching down.*
Right *The remarkable looking Lunokhod.*

THE SPACE AGE 83

Driving the machine by remote control from Earth must have been a trying experience for the five-man team, since there was a delay of three seconds between moving a control and seeing what happened. The results produced by Lunokhod 1 were most satisfactory and gave some substance to the Soviet comments on the dangers of manned exploration.

Without developing a really large launch vehicle to match the American Saturn V, the Soviet space programme could advance no further. Since 1967 there had been rumours in the West that work was being done on a giant vehicle, and James Webb, Administrator of NASA, used to frighten Congress with it every year at budget time. The craft was nicknamed 'Webb's Giant' in his honour, and photos taken by satellite were studied closely to see if it existed. A launch-pad was discovered with unusual activity on it. But on the next pass by the satellite the rocket that had been on the pad—and the entire pad itself—were gone. Blown up. Later tests seemed to have ended in failure as well. Despite this there is no doubt that some day the Soviets will develop a lifting body in the Saturn V category.

Lack of a craft this size was still holding back their space programme. Proton was just big enough to launch their first orbiting space station, Salyut 1, on 19 April 1971. It weighed almost 19 tons and contained two main areas for living and working, cylindrical in shape, 9 feet in diameter and 25 feet long. When completed it also had a Soyuz ship built into it. Though it was meant to be a permanent habitat in space, its low orbit caused it to fall back to Earth within six months. Lack of a really big booster meant that the low-powered

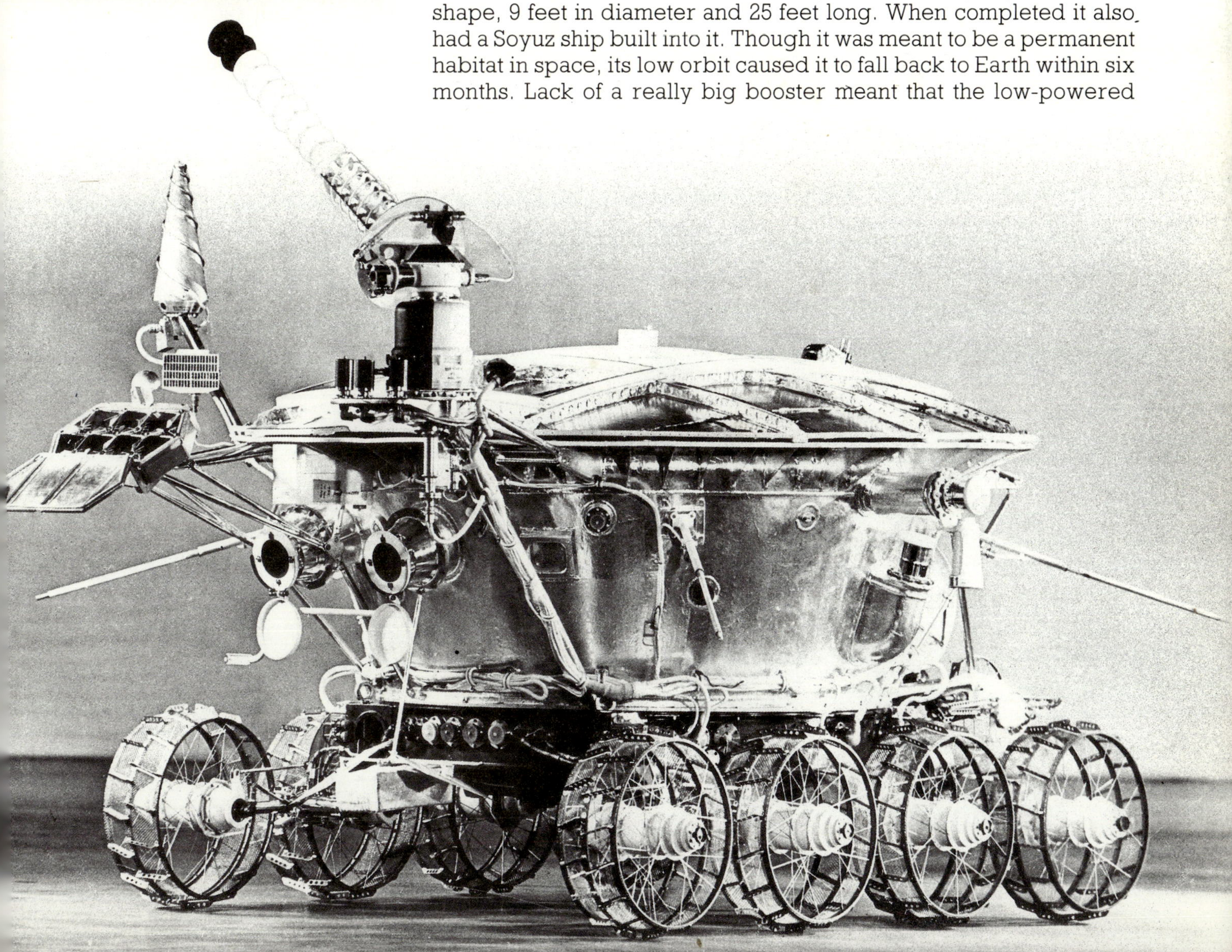

84 THE SPACE AGE

Right *Skylab crewman during extravehicular activity.*
Below *Skylab in orbit. The flag-like projection is the single operative solar panel, its twin having been torn away.*

Soyuz could not get a three-man crew any higher. Soyuz 11 entered Salyut on 6 June 1971 and the cosmonauts spent 23 days working aboard. This successful mission ended in tragedy when, during re-entry, a faulty valve caused the craft to depressurize and the crew suffocated. Spacesuits could not be worn because of the cramped conditions, just as in Voskhod. Later missions carried only two men so there was room for their spacesuits.

More unsuccessful launches followed the demise of Salyut 1, and it was not until June 1974 that Salyut 3 was put into a higher and better orbit and remained there for two years.

Not to be outdone, the Americans got into the space station business in a big way with Skylab which was put into orbit in May 1973. It was far bigger than the Soviet lab, a cylinder 47 feet long and 21 across, weighing over 75 tons. More than enough room for a three-man crew to live and work in. The only trouble was that vibrations during launching had caused serious damage. Its thermal shield and one of the two solar panels had been torn away, while the remaining panel was jammed in a partially open position. Without the thermal shield heat would build up quickly and bake anyone inside. However, in the true tradition of the fictional spacelanes now turned to fact, the first Skylab crew went up anyway in their Apollo craft. They managed to open out a specially built umbrella that would shield Skylab from the sun's rays, then struggled open the jammed

A Soyuz spacecraft approaches a Salyut space station. Near the nose of each craft are matched pairs of antennae which ensure proper alignment during docking.

86 THE SPACE AGE

Left *An unmanned Russian probe lands on the burning surface of Venus.*
Below *A Mariner spacecraft was the first operational sunjammer, using the pressure of light on its solar panels to make course corrections.*

panel. This proved to be more than adequate to supply all the electricity needed so the crew moved in. There were three manned trips to Skylab and each set a new space endurance record. The final one, set by Carr, Gibson and Pogue, lasted for 1,214 orbits, 84 days.

Skylab marked the end of America's manned space programme for the time being. It also proved to be more than a slight embarrassment since it was slowly falling back to Earth. Both America and the Soviet Union have been less than truthful when they have stated that space debris falling back to Earth burns up on re-entry. It certainly does not. Entire great rocket boosters have been sighted floating in the ocean after having allegedly burned up. What both countries have done is bank upon the law of averages. Three-quarters of the globe is covered with water, while only a small percentage of the total land-mass is inhabited. So until now almost all space garbage has landed in the ocean, or hit the desert or mountains and thus no one has noticed. But people certainly did notice when a Soviet satellite spread radioactive debris over Canada, and they will notice it even more when 75 tons of metal drop on their heads. There were plans for the new American Space Shuttle to boost Skylab into a higher orbit, but it won't be ready to fly in time. Fiction has antedated fact again, because Skylab is falling for the same reason that disaster threatens a satellite in Harrison's 1976 novel, *Skyfall*. Solar storms increase the radiation from the sun. This

The Voyager space probe will reach Saturn in 1980, having already transmitted the most remarkable pictures of Jupiter ever seen.

causes the top of the atmosphere to warm up and rise abruptly. The increased friction of the air causes the satellite to slow down and fall. The chances are that NASA will win the toss of the dice again, the odds are so good, and that Skylab will hit water or wasteland. But it will come down and it still weighs 75 tons—not a pleasant prospect.

The future of manned spaceflight cannot be predicted with any certainty beyond the early 1980s. Lack of money has forced NASA to shelve the plans for landing on Mars in the next decade, though they may be dusted off for use in the 1990s. Unmanned flights though are proceeding apace in both the United States and the USSR. Though they do not have the drama of manned spaceflight, they are scientifically of the greatest importance. Soft landings on Venus and Mars have sent back incredibly vital data, as has the fly-by of the outer planets.

While the Soviets have little to say about future manned spaceflight, the American Space Shuttle will soon be in regular service. This is the first reusable space vehicle and is much more in line with all of the fictional and factual predictions of the pre-Space Age era, none of which visualized the present practice of expensive hardware being discarded like empty beer cans after one use. The Shuttle, about the size of a jet airliner, is launched vertically by two large solid fuel boosters, with the help of its own engines fuelled by a large external tank. The boosters drop back by parachute to be used

88 THE SPACE AGE

again, but the tank is thrown away in the ocean to conform to normal practice. Well on its way, the Orbiter part of the Shuttle goes into orbit. Not only does it take up a payload of 30 tons, but it can bring half that weight back to Earth. Upon its return its aerodynamic shape lets it descend like a great glider and drift down for a deadstick landing. Since all the major components are reusable, NASA is looking forward to a schedule of routine operations and is already selling space on the first flights. Sixty flights a year are projected for the 1980s. One of the first things it will do is carry the European Space Agency's Spacelab up into orbit. Largely funded by West Germany, Spacelab—a successor to Skylab which will go into permanent orbit—is Europe's only significant contribution to date to the manned space programme. Among its crews will be Western Europe's first astronauts who are currently in training. Personnel will be ferried back and forth by the Shuttle, and scientific and commercial research projects are already prepared for the day. Science fiction not only predicted a shuttle like this to service satellites, but has even named the first Shuttle Orbiter *Enterprise* following a postal campaign by thousands of *Star Trek* fans.

A more ambitious project that is slowly gaining support, though no official action is yet planned, is the building of permanent

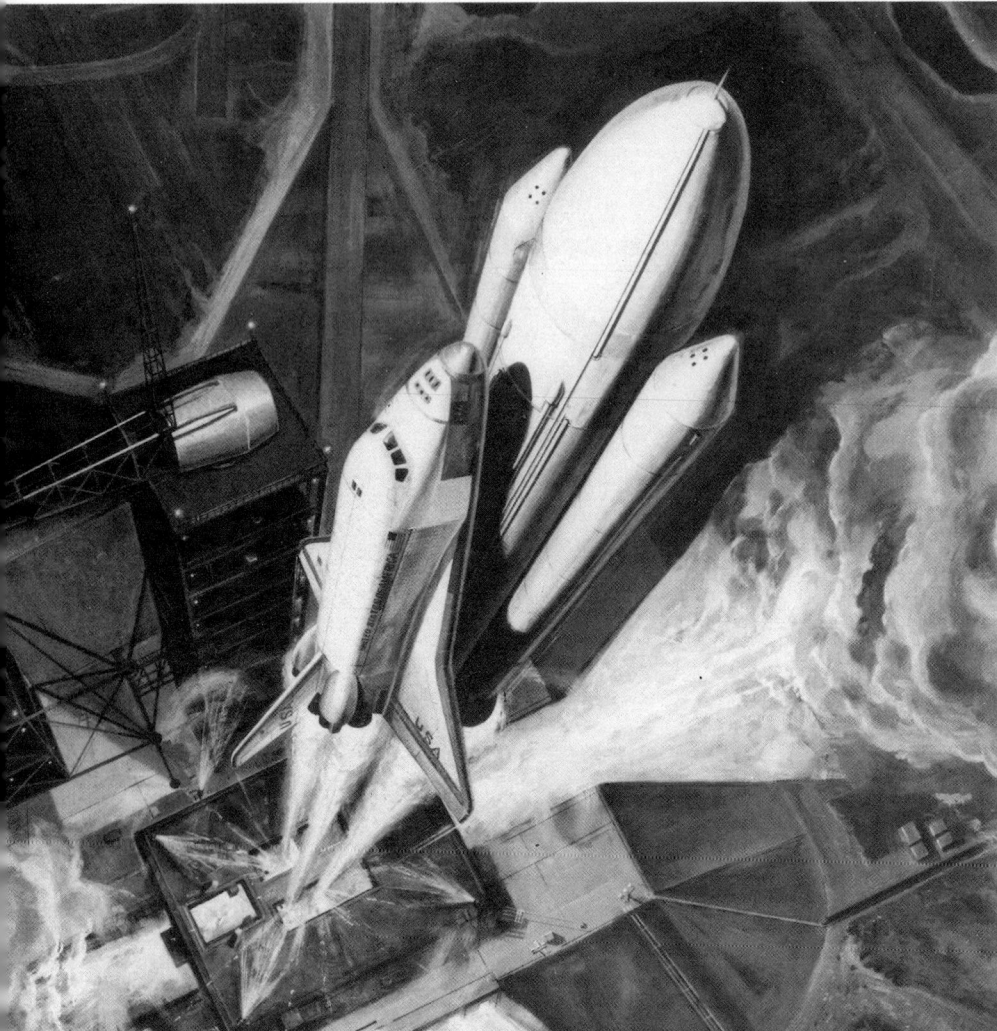

Left *The Space Shuttle is launched with the assistance of strap-on boosters and fuel tank.*
Above *The Shuttle in orbit with the European Space Agency's Spacelab shown in its payload bay.*
Right *Gliding in to touchdown after a flight test.*

habitats in space. These space colonies have been designed in great detail by Professor Gerard O'Neill of Princeton University, although earlier descriptions of colonies can be found in the writings of Tsiolkovsky and Bernal as early as the 1920s. This is Bernal's view:

Imagine a spherical shell ten miles or so in diameter, made of the lightest materials and mostly hollow. Owing to the absence of gravitation its construction would not be an engineering feat of any magnitude. The globe would fulfil all the functions by which our Earth manages to support life. In default of a gravitational field it has, perforce, to keep its atmosphere and the greater portion of its life inside.

The outer shell would be hard, transparent and thin. Its chief functions would be to prevent the escape of gases from the interior, to preserve the rigidity of the structure, and to allow the free access of radiant energy. Immediately underneath this epidermis would be the apparatus for utilizing this energy either in the form of a network of vessels carrying a chlorophyll-like fluid or some purely electrical contrivance. In the latter case the globe would almost certainly be supplied with vast, tenuous, membranous wings which would increase its area of utilization of sunlight.

Bernal's colonists spend their time in free-fall inside his globe, since he did not rotate it to provide artificial gravity. Other than this it is a remarkable prediction and very close to current design studies that are being done. An early science fiction use of the idea occurs in Jack Williamson's 'The Prince of Space' (1931).

The present work dates to 1974 when O'Neill first wrote about his ideas in *Physics Today*, then spoke to a small conference at Princeton. Interest snowballed rapidly after this and even NASA came up with some money for research. Several groups came into existence to boost and publicize space colonies. The most important of these is the L-5 Society, named for one of the locations in the Earth-Moon system—first predicted by the French mathematician, Lagrange—where any object will remain in permanent stable equilibrium. O'Neill's first idea was for a rotating cylinder that would provide gravity for people living on the inner surface. It would be a mile long and 600 feet in diameter. This is far larger than anything else ever put forward as a plan for the space programme, but that is the whole idea of the project. It must be a space *colony*, and appear

Bernal Sphere design for a space colony. Rings of huge mirrors reflect sunlight into the interior.

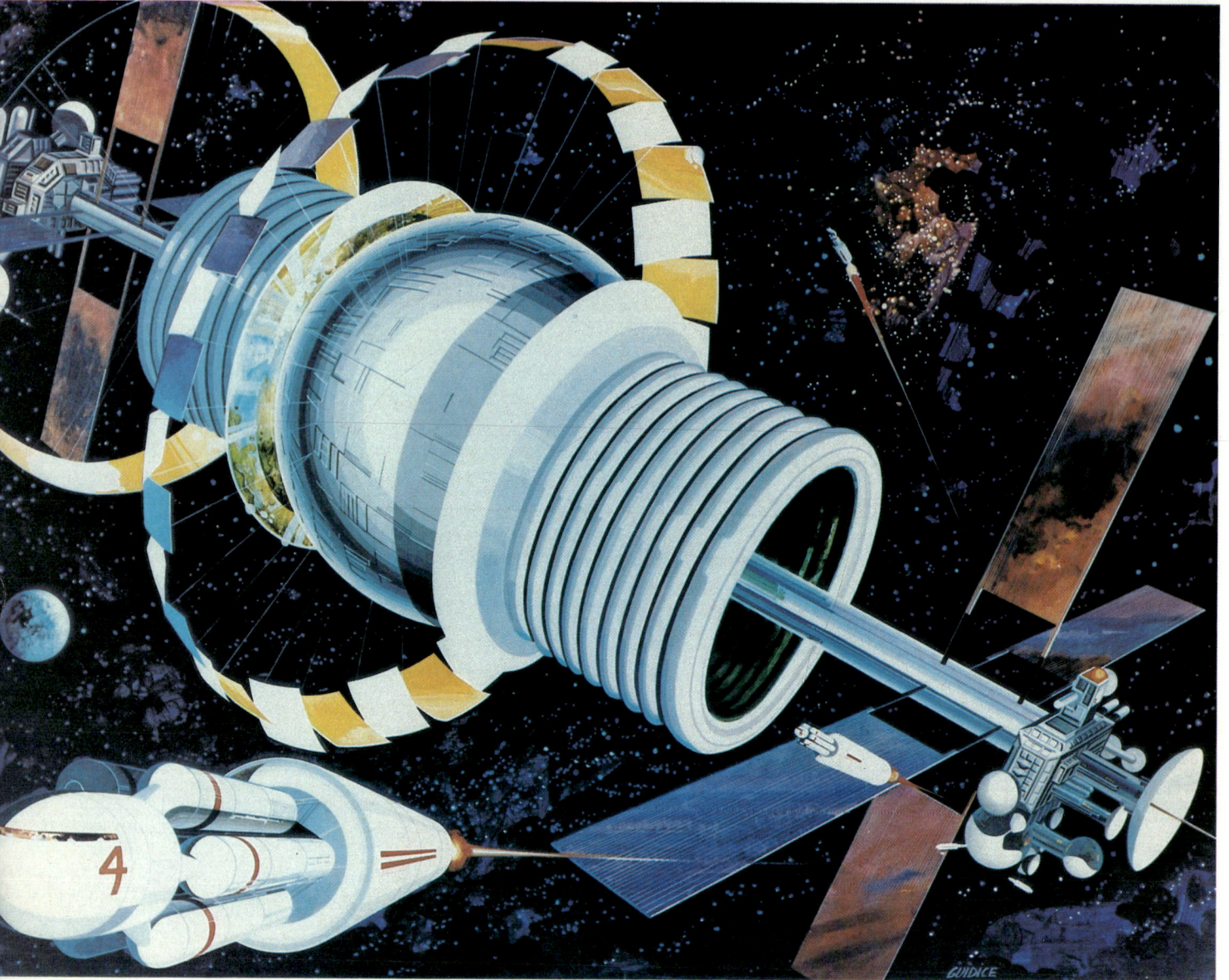

Cutaway of Stanford torus colony, showing the spacious and comfortable environment. Glass windows overhead admit sunlight.

more like an entire world than a spaceship to its inhabitants. Not the world we are familiar with, obviously, with people and buildings hanging over our heads, but still a world. Certainly an environment like this would produce at least one new phobia for the textbooks.

One difficulty with the original design was that it would have to rotate at a considerable speed to simulate normal gravity, and this rotation would have some uncomfortable side-effects. If anything were dropped, for instance, it would not fall straight but would instead whip off to one side. O'Neill has done some rethinking on the design and now favours a sphere, christened the Bernal Sphere in honour of the man who first described it. Another design is the Stanford torus, a doughnut-shaped colony that resembles a blown-up version of von Braun's wheel in space. It differs in size, so instead of being divided into compartments it has a single open space inside, like the innertube of a tyre. One advantage is that the inhabitants would not have to look up at others suspended over their heads.

Although a space colony would be a massive engineering project, it could be done with today's technology. But transporting all the raw material up from Earth would be incredibly expensive

92 THE SPACE AGE

NASA design for a space station which can be assembled in orbit from modules prefabricated on Earth and carried up by the Space Shuttle.

even with special space shuttles modified as pure cargo carriers. The design projections therefore include a mining colony on the moon where all of the raw materials exist for basic construction of the satellite colony. Spacecraft will not be needed to transport the material to the construction site—it will instead be 'shot' there in packages. The low lunar gravity, the airlessness and the low temperature make possible the construction of a magnetic linear accelerator. Supported and moved by magnetism the bundles will be accelerated along a track and shot out into space—where they will be captured by a giant funnel-shaped net at journey's end. This sounds like a fantastic science fiction plan—but it is immensely practical.

Even more practical is the power supply for the colony. Solar power, free energy forever. Sunlight falls on solar cells which generate electricity, or the sun's heat vaporizes a liquid in sealed tubes which runs a conventional generator. There is a greater need for a generating satellite than there is for a colony and we may be seeing them constructed in the very near future.

The world is running out of energy. We all know this in a fairly general way but the reality never quite strikes through. It will when all of the oil runs out in about twenty years. Now is the time to begin planning for that possibility, and solar energy is one of the most promising alternative energy sources. A solar generator would be in a geosynchronous orbit, in other words it is always above the same spot on the surface below. Light would be turned to electricity, which would be broadcast towards the Earth as short-wave radio waves. Special arrays of antennae would receive the short-waves and turn them back into electricity to feed the grid. Practical, possible and a little expensive. But still far less than any military budget. And once the first one is built the others will be free—paid for by selling the electricity generated in space.

With this solar power the first colony—Island One—can be built. While its construction is still a matter of doubt we can be sure that there will be numerous science fiction stories telling us all about it; Ben Bova's 1978 novel *Colony* is among the first. However, the feasibility studies from respectable universities and government institutions show concepts that would have been science fiction a few years ago but which have now become established as part of mankind's possible future.

There is another possible use for space colonies that has been completely ignored. After the first few generations of life in a colony it will be seen as home; Earth will be as China is to a Westerner now. Undeniably there, of a certain interest, but never to be visited by the overwhelming majority. All one's relatives and friends will be in the colony and life will be complete and pleasurable. A perfect recruiting ground for the 10,000 madmen needed to populate a space ark! At last the perfect citizens for a colony ship. With fusion power taking the place of solar power the colony would be ready to make the journey between the stars. No worry about the breakdown of machinery or morale because all of these problems would have been faced and solved long before. So hollow out your asteroid, stock and furnish it and let the happy throngs troop aboard. We're off to the stars!

Above *A mining colony on the Moon. To the left is the track of the linear accelerator which will shoot raw materials off into space.*
Left *A gigantic solar power station in Earth orbit.*

TRAVELLERS TO THE STARS

Our sun is a star like all of the others, so if we are planning to visit suns out there—perhaps they are planning to visit us as well. If we can build space arks to undertake millennia-long journeys, then other civilizations (some of whom, by the law of averages, must have been around a lot longer than *Homo sapiens*) may long ago have set off in our direction. Science fiction gives us a description of just what one of these alien interstellar arks might look like, a really large and disorientating one for the humans who explore it:

Safest of all was to imagine that he was at the bowl-shaped bottom of a gigantic well, sixteen kilometres wide and fifty deep. The advantage of this image was that there could be no danger of falling further; nevertheless, it had some serious defects.

He could pretend that the scattered towns and cities, and the differently coloured and textured areas, were all securely fixed to the towering walls. The various complex structures that could be seen hanging from the dome overhead were perhaps no more disconcerting than the pendent candelabra in some great concert-hall on Earth. What was quite unacceptable was the Cylindrical Sea . . .

There it was, half-way up the well-shaft—a band of water, wrapped completely round it, with no visible means of support. There could be no doubt that it *was* water; it was a vivid blue, flecked with brilliant sparkles from the few remaining ice-floes. But a vertical sea forming a complete circle twenty kilometres up in the sky was such an unsettling phenomenon that after a while he began to seek an alternative.

That was when his mind switched the scene through ninety degrees. Instantly, the deep well became a long tunnel, capped at either end.

This is from Arthur C. Clarke's *Rendezvous with Rama* which was published in 1973, well before O'Neill's first article on the subject. Again science fiction was there first.

One fantastic idea for interstellar travel was put forward by Robert W. Bussard in 1960 in a paper published in *Astronautica Acta*. It was titled 'Galactic Matter and Interstellar Flight' and, though it sounds like the title of a bad science fiction story, it is a serious scientific proposal to solve this problem. Bussard suggested that a certain kind of ship could be built that would use the finely dispersed material in interstellar space as fuel. There are hydrogen and helium atoms out there, not very many—about a molecule of matter per

Fusion-powered attack vessels circle an unwieldy nuclear cargo ship—modern SF hardware at its best, by Peter Jones.

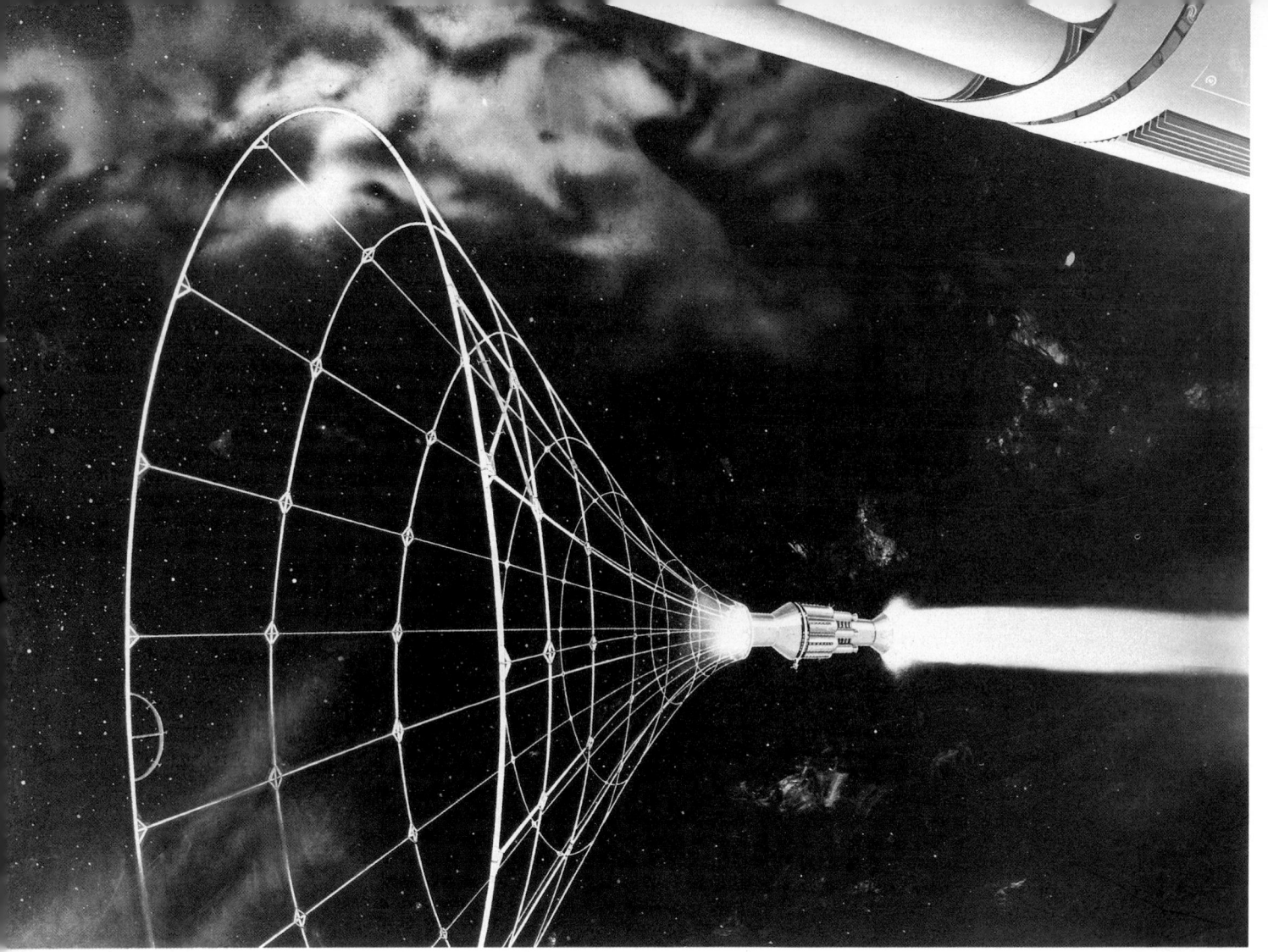

A Bussard ramjet, fusion torch blasting at full power, races through interstellar space.

cubic yard—but still enough if you went about gathering them in the right way. A spacecraft moving at very high speed would scoop up enough of them to use as fuel, a sort of interstellar ramjet. The scoop would have to be very large, nearly 2,000 miles across, which would present certain problems of construction. A solid structure of such size is not feasible, but it would be possible to generate a magnetic field which would scoop up and channel the ionized gas into the engine's maw. It would have to be moving very fast to get enough fuel to keep it going, and would get 50 per cent thrust at about 2 per cent of the speed of light. Sounds reasonable, a mere 3,720 miles a second, and a fusion engine might be developed to boost it to that speed so the ramjet action could take over. At journey's end the magnetic field would be reversed to repel rather than draw in the gas, creating a drag to slow the ship down. Bussard's idea was quickly seized by the science fiction writers, who know a good thing when they see it, and first off the mark was Poul Anderson in 1967 with 'To Outlive Eternity', later novelized as *Tau Zero.* Of course something has to go wrong—or there would be no story—and his ship goes out of control when the Bussard drive can't be cut off. Their speed builds up closer and closer to the speed of light and, Einstein notwithstanding, relative time in the ship slows down. They go bumping and rattling through entire galaxies and, as the author puts it, 'They talked business for half an hour. (Centuries passed beyond the hull.)'

Larry Niven also writes of ramscoops, though without this high drama, and here is a description of one from his 1968 *A Gift from Earth*:

Interstellar Ramscoop Robot # 143 left Juno at the end of a linear accelerator. Coasting toward interstellar space, she looked like a huge metal insect, makeshift and hastily built. Yet, except for the contents of her cargo pad, she was identical to the last forty of her predecessors. Her nose was the ramscoop generator, a massive, heavily armored cylinder with a large orifice in the center. Along the sides were two big fusion motors, aimed ten degrees outward, mounted on oddly jointed metal structures like the folded legs of a praying mantis. The hull was small, containing only a computer and an insystem fuel tank . . .
  The ramscoop generator came on.
  The conical field formed rather slowly, but when it had stopped oscillating it was two hundred miles across. The ship began to drag a little, a very little, as the cone scooped up interstellar dust and hydrogen. She was still accelerating. Her insystem tank was idle now, and would be for the next twelve years.
  Her food would be the thin stuff between the stars.

Another form of propulsion that has been suggested is a nuclear pulse rocket. Behind this imposing title lurks the spirit of the mad design of our old friend Ganswindt. He had planned to shoot fragments of shrapnel from TNT bombs to get propulsion. While some might hesitate before taking a trip in this flying-bomb rocket, more would think twice about riding a nuclear pulse rocket which is propelled by a series of atomic bombs going off in the tail. Yet this idea was seriously examined by Project Orion in the U.S.A. in the 1950s. But fission bombs are dirty things that would emit a highly

Inside the gigantic alien spaceship Rama.
Overleaf *Chains of curious alien spacecraft encircle the Earth in this dramatic Tim White painting.*

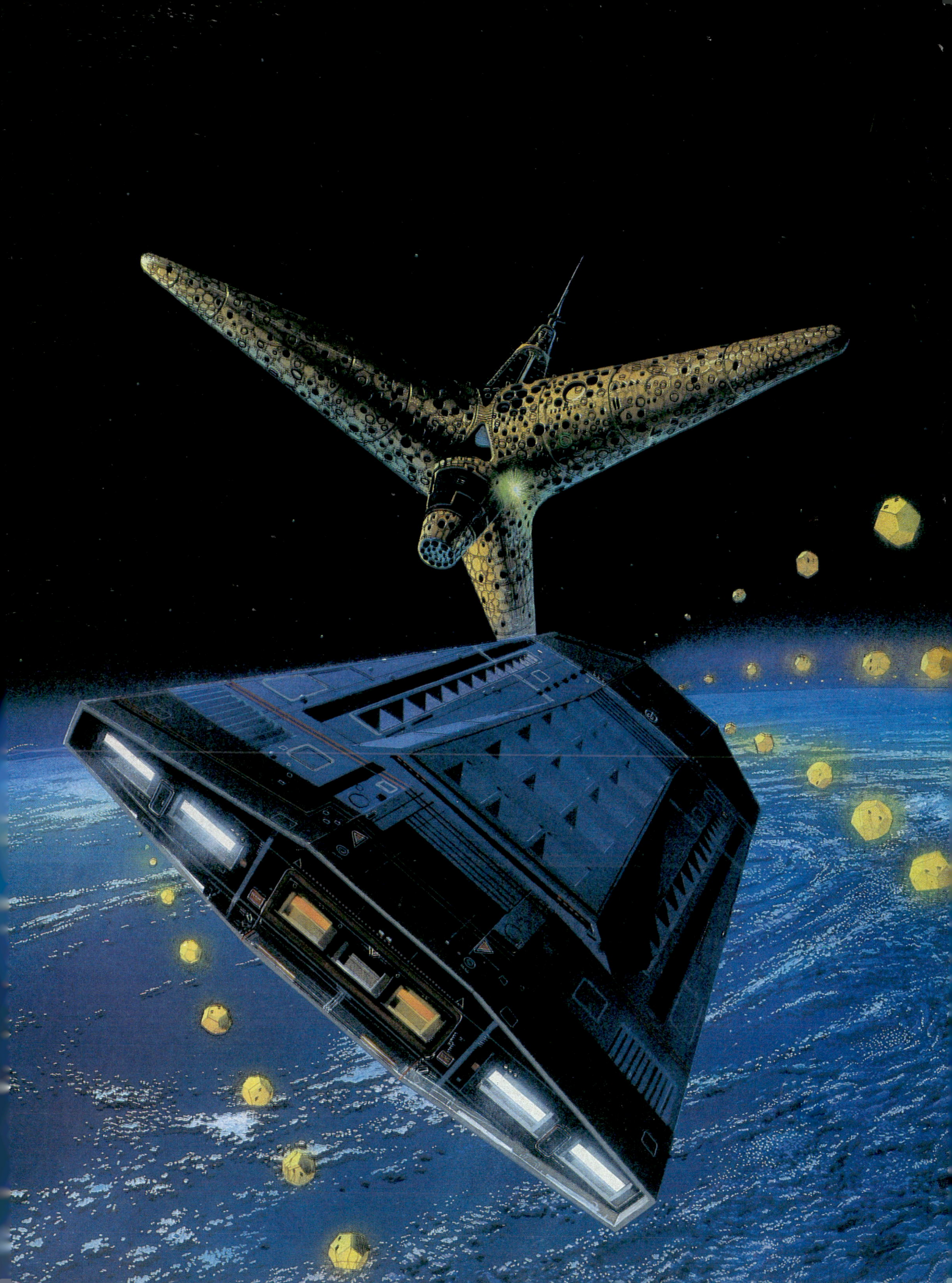

radioactive exhaust. Fusion bombs are much cleaner, at least in this usage, and more efficient at energy production. Freeman Dyson designed—in theory—a fusion-pulse rocket that would be propelled by one-megaton H-bombs, one of them going boom every three seconds. It would take 300,000 of the bombs to move the ship at an acceleration of one G for ten days. At which point the craft would be going at 3 per cent of the speed of light and the engine could be shut down and the crew could take out their earplugs. This nuclear pulse rocket would be able to carry a payload of 5,000 tons.

It is pleasant to know that the British Interplanetary Society is still in business, and still prevented from experimenting by the British Explosives Act. So it is back to the drawing board and, as they designed a gunpowder rocket before, they are now at work on the design for a nuclear pulse rocket. Their starship feasibility study is called Project Daedalus. They have produced a handsome thick volume of design, complete with equations, for detonating a series of very small fusion reactions by means of focused beams of electrons. The fuel is a mixture of Deuterium and Helium 3 compacted into small fuel pellets. These will be bombarded inside a reaction chamber and a magnetic field will channel the resulting plasma through the

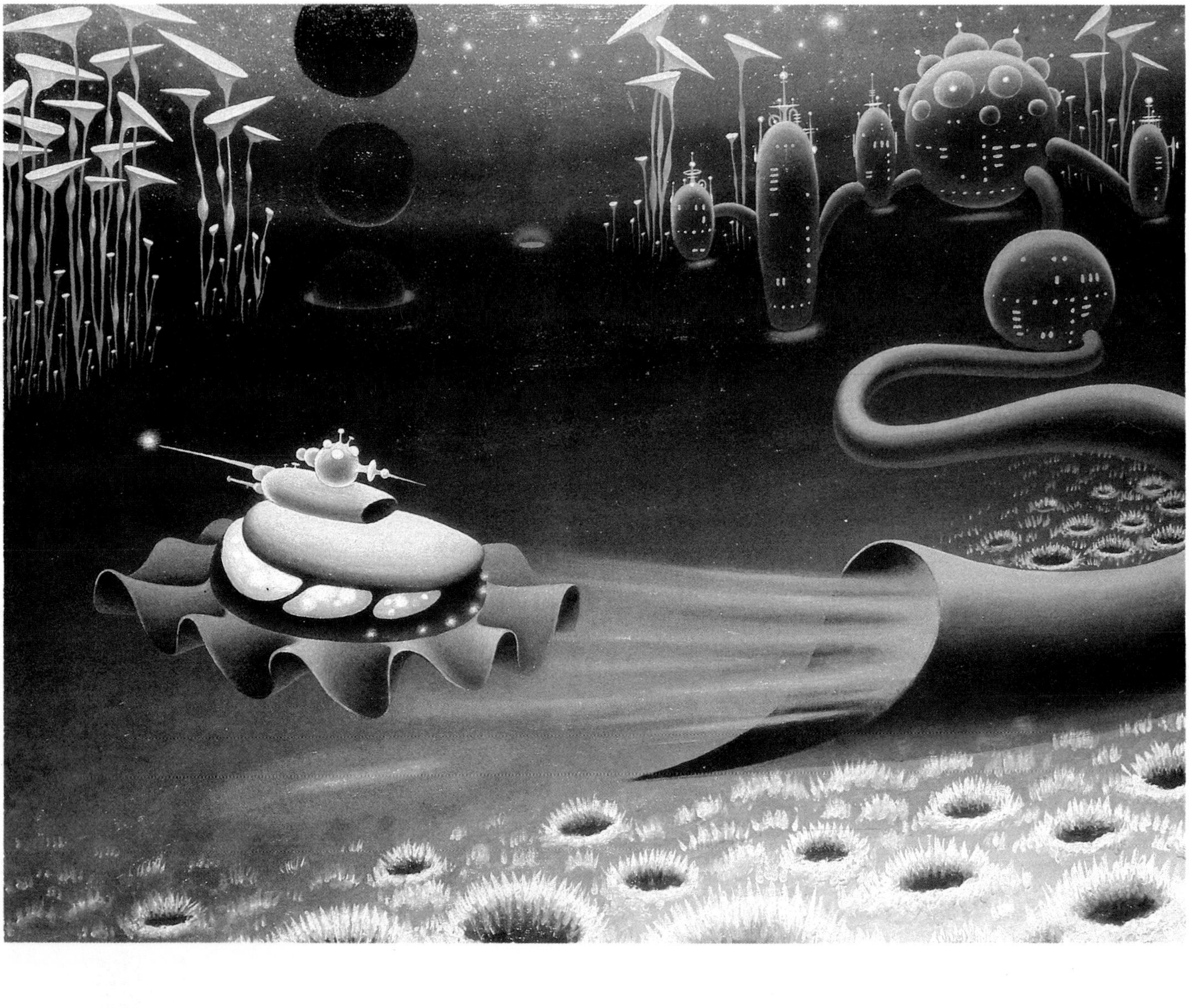

TRAVELLERS TO THE STARS 101

exhaust to provide thrust. Since there would be 250 of these explosions each second, the ride should be rather smoother than Freeman Dyson's design which has one enormous one going off every three seconds.

There is one problem about the fuel though; there is not enough Helium 3 on Earth to fill the bunkers. Therefore the first leg of the interstellar voyage is within our own solar system—a fuelling stop at Jupiter. There is plenty of this rare gas in Jupiter's frigid soup of an atmosphere so the first part of the Daedalus mission will involve building orbital factories to scoop up as much helium as is required. Once the tanks are full the big bang-bang begins and the ship is off in the direction of Barnard's Star, building up to its top speed which should be close to 13 per cent of the speed of light.

Daedalus is the most ambitious and detailed proposal yet put forward for an interstellar voyage. The probe would be unmanned; after all, Barnard's Star is 5.91 light years away, and the trip would take about fifty years to complete. But there would be six to nine years of radio transmission time at the other end when the probe would send back detailed information. It would take at least twenty years to design and manufacture this craft. Nor will the ship be built

Science fiction art from behind the Iron Curtain.
Left *A petalled spaceship is launched underwater: a surreal vision by N. Nedbailo.*
Above *A classic flying saucer design by Russia's leading space artist, Andrei Sokolov.*
Right *Cosmonauts working in free fall secure themselves to the ship's hull with clumsy magnetic boots in this Sokolov painting.*
Overleaf *Next stop Barnard's Star as Project* Daedalus *blasts off from Jupiter orbit.*

Ion-powered rockets being assembled in orbit: a typically realistic Bonestell illustration.

tomorrow. It is estimated that construction cannot begin until the late twenty-first century by which time, it is hoped, the technology will be available to mine Jupiter's atmosphere. So this is really a more conservative and long-range study than the one done by the British Interplanetary Society in 1939, when they said that a moonship could be built there and then if only the funds were available. One of the members of the present design team, Gerry Webb, reports that Project Daedalus is meeting precisely the opposite reaction to that of the first rocket study. The first one was thought to be a kooky idea put together by a bunch of nuts and was not taken seriously. Now, in a world of moon rockets, atom bombs, and *Star Trek* fans, the job is to convince people how difficult the project will be. They expect the thing to be knocked together quickly, the switches pulled—and Barnard's Star next stop.

Yet there is plenty to see right here in our own solar system, before an interstellar flight is planned, and certainly there must be improvements possible on today's liquid fuel rockets. One such drive could be the ion rocket, where a nuclear reactor is used as power to send a stream of ions jetting out the stern. Ionized particles are not very heavy, but throw enough of them away fast enough and you get thrust. Once again J. D. Bernal anticipated the problem and suggested at least one possible solution:

To set in motion any large rocket, the mass of gas required is of the same order as the weight of the rocket itself, so that it is difficult to imagine how the rocket could contain enough material to maintain its propulsion for any length of time. When the radio-transmission of energy is effected half the difficulty will be removed, and the projection may very well ultimately be effected by means of positive rays of high potential.

TRAVELLERS TO THE STARS 105

For 'positive rays of high potential' read lasers and you are on the way to a solution. Since a sunjammer can move in only one direction, away from the sun, another power source must be found for its sails. Dr Robert Forward suggested that a group of lasers be put into orbit, deriving their power from the sun. As usual, as soon as a theory is proposed the science fiction writers spring forward and build the hardware to launch it into motion. Niven and Pournelle leapt first in 1974 with *The Mote in God's Eye*, in which an alien sunjammer has been discovered approaching from a mysterious area of space known as the Mote. But one of the characters has the answer:

'Captain, look,' he said, and threw a plot of the local stellar region on the screen. 'The intruder came from here. Whoever launched it fired a laser cannon—probably a whole mess of them on asteroids, with mirrors to focus them—for about forty-five years, so the intruder would have a beam to travel on. The beam and the intruder both came straight in from the Mote.'

Of course none of these fuel problems arise if you make your spaceship big enough—that is, use the spaceship we already have. Earth. To think is to act, and a number of mobile worlds have whizzed through the pages of science fiction. In Donald Moffitt's *The Jupiter Theft* (1977) the aliens come thundering into our own system in their space colonies towing an entire planet with them. It seems that they

Rick Sternbach is a young American artist firmly in the Bonestell tradition.
Below left *An interstellar probe undergoes repair in Earth orbit.*
Below right *Multi-national sail-powered spacecraft approach Mars.*

106 TRAVELLERS TO THE STARS

use giant gas planets as fuel for their drives, stripping them of their hydrogen. When they enter a new solar system they throw away the bare rock core of the used-up planet and pinch a new one—explaining the title of this novel.

In 1964 Fritz Leiber had another planet enter our solar system in *The Wanderer*. This is what it looked like:

> Gradually the three-dimensional picture firmed in his mind of the Wanderer, artificial throughout, globe within globe of floors—fifty thousand of them at least—everywhere veined with corridors, like a vast silvery sponge. Many of the great wells did go all the way through the planet, intersecting at its center in an immense empty globe that had a dark sky of its own glittering with random lights like stars between the mile-wide holes of the pits with their darkness and their softly glimmering lights.

Each of these plans is a way of getting around the speed of light. It was so much easier in the good old days with the FTL drive. It lived in a black box and you actuated it by pressing a button. Zingo!—and there you were. But that is more magic than science and we like to feel that things have progressed a bit in SF. The tachyon theory helps, for if you can postulate particles that can travel only faster than

Left *Nothing can escape the vicinity of a black hole. Rocks, small planets, any unfortunate spaceships—all are sucked in. Tony Roberts paints the scene.*
Above *A Chris Moore space machine slides ominously over a jagged futuristic landscape.*

the speed of light, you can certainly postulate a tachyon drive.

Another up-to-date theory that is getting strong attention in the pages of SF is the theorized existence of black holes. As astronomical evidence piles up to lend credence to the theory, it gets more and more attention from the authors. A black hole is what remains of a giant star with at least three times the mass of our own sun. The theory postulates that as these suns grow older and cooler they collapse upon themselves because of the immense forces of gravity. As they get smaller the gravity is concentrated and eventually becomes so strong that even light rays cannot escape its pull; hence the name of the theory. Eventually it reaches a point

108 TRAVELLERS TO THE STARS

where its diameter approaches zero and its mass is heading towards infinity. If this black hole is rotating, and latest evidence seems to suggest that this can happen, centrifugal forces would pull it into the form of a flattened disc. At the edge of this disc, where normal space meets the warped space of the black hole, there would be an area of transition. Again it is theorized that if anything passed through this area it would be instantly transported to a different part of the galaxy. Statements like this are grist for the SF mills. In Joe Haldeman's *The Forever War* (1974) the black holes are called 'collapsars' and are used for the interstellar travel needed to keep the war going. In Adrian Berry's *The Iron Sun* (1977) the serious proposal is made that not only could black holes be used for interstellar travel, but also that they be created artificially for this purpose.

If you can't build your own starship you can always borrow one from the aliens. This not only makes for an interesting plot but

Heechee ships set off from Gateway *into the unknown: Boris Vallejo illustrates Frederik Pohl's novel.*

An immense floating city hovers above the Earth—a modern rendition of one of SF's most enduring visual images by Robert McCall.

lets you get around the explanation of how the thing works. In Frederik Pohl's *Gateway* (1977) a convenient cache of long-abandoned alien spaceships is found on an artificial asteroid. They can go faster than light—but no one can figure out how they work. You just get aboard, press the button and hope that the programmed instructions will take you to some place interesting. Nor are they very large:

The inside of a Heechee ship, even a Five, is not much bigger than an apartment kitchen. The lander gives you a little extra space—add on the equivalent of a fair-sized closet—but, on the outleg at least, that's usually filled with supplies and equipment. And from that total available cubage, say forty-two or forty-three cubic meters, subtract what else goes into it besides me and thee and the other prospectors.

Some people return from journeys in a Heechee ship. Some don't. And thereby hangs a plot.

With the growing interest in science fiction books and films we no longer have to use our imaginations to know what a spaceship looks like. Frank R. Paul broke the ground and many talented artists now follow after him. For far too long artists slavishly copied each

110 TRAVELLERS TO THE STARS

The awesome scale of a Foss spacecraft as experienced through the eyes of approaching astronauts: illustration for John Varley's Ophiuchi Hotline.

other's designs so that Standard Issue Rockets flew identically from magazine covers and book jackets. Now this has all changed—for the better—and the new generation of space artists is perhaps best represented by Chris Foss, whose first book covers appeared in 1970. Foss's original training as an architect can be seen in the muscular solidity of his designs. His creations look as though they could work. His spaceships come in an immense variety of shapes and sizes, with an illusion of imposing reality generated by the dense surface detail. Foss says of his work:

TRAVELLERS TO THE STARS 111

I paint science fact not science fiction. My work shows a time that is only one step above life today. I take things happening today and add on about 150 years... I pick up basic principles—how things work—and quite often they come back to me when I need them for a cover. Things I see or talk about are often useful later. Working on science fiction you find that mechanical principles similar to those today are going to be used in the future.

What Foss does not say is that he is also a dramatic painter, not a simple illustrator. In addition to the artistic excellence of his work

112 TRAVELLERS TO THE STARS

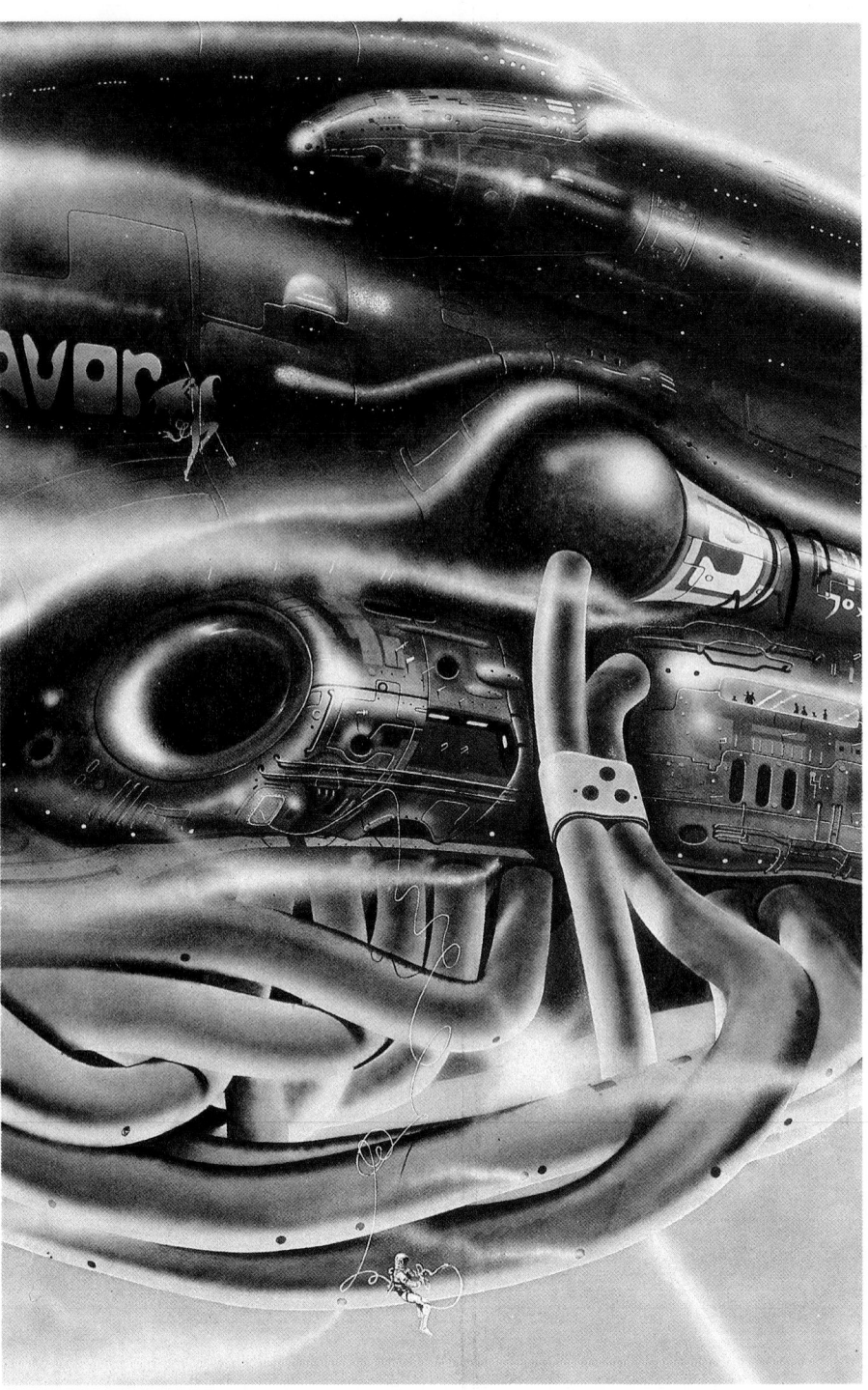

This Jim Burns spaceship, with its twisted mass of intestine, clearly has a life of its own.

there is the fact that something is always happening in his art. A perfect example is the sinister metallic arm that thrusts up from the water to clutch the passing spaceship. This must have been his idea—it can't have been in the text he was illustrating.

Foss's success has fostered a whole school of British SF artists, all of whom employ his smooth airbrush techniques and crystal-clear neo-realism. These include Anthony Roberts, Angus McKie and the early work of Peter Jones. Equally realistic is the work of Jim Burns, who is nevertheless a complete and immensely satisfying artist in his own right. Burns's spaceships and machinery are warmly organic, possessed of life. Sometimes directly, many times by suggestion, Burns embodies the Freudian suggestion of spaceship as phallic symbol in his work.

In the United States the leading spaceship artist is perhaps Rick Sternbach, who specializes in realistic spacecraft covers for magazines. His craft are almost always extensions of existing technology, for the current American approach is very representational. Not so in Britain, for it is doubtful if even Foss himself could prepare blueprints for ninety per cent of his craft. But in the United States some of the best artists are employed directly by NASA to prepare official scenarios illustrating possible future developments of the space programme. Don Davis has painted many of the space colony visualizations, while Robert McCall has done vast and impressive—and realistic—murals to decorate the new Air and Space Museum in Washington. These artists are the direct descendants of Chesley Bonestell. In this tradition, just as Bonestell did paintings for *Destination Moon* so has McCall designed spaceships for a Disney film.

Film is very important in current spaceship design and the film-makers have broken new ground—or new space. It all began with *2001: A Space Odyssey* and shows no sign of stopping or even slowing down yet.

A mile-long spaceship designed by Robert McCall for Disney's Black Hole.

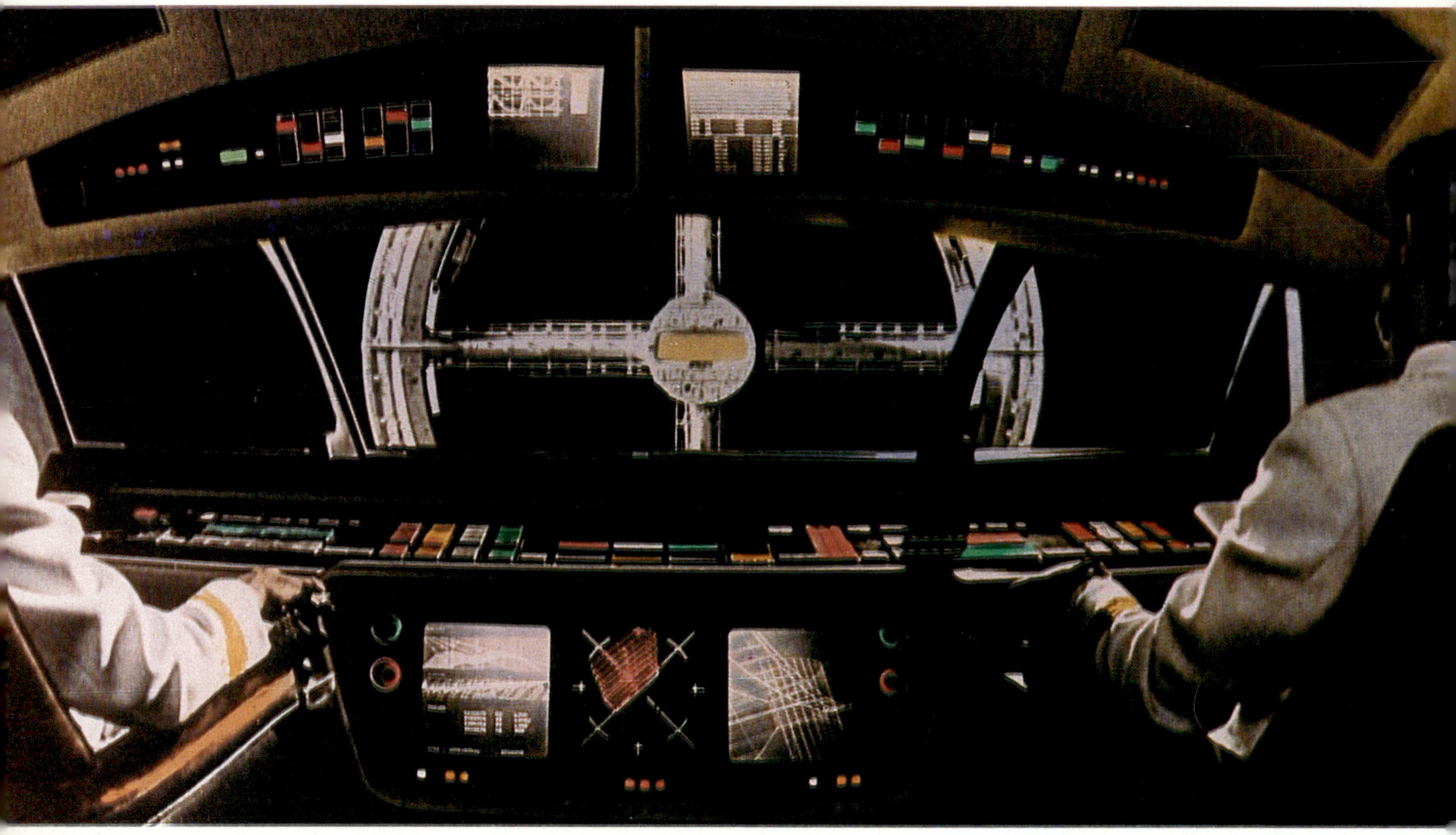

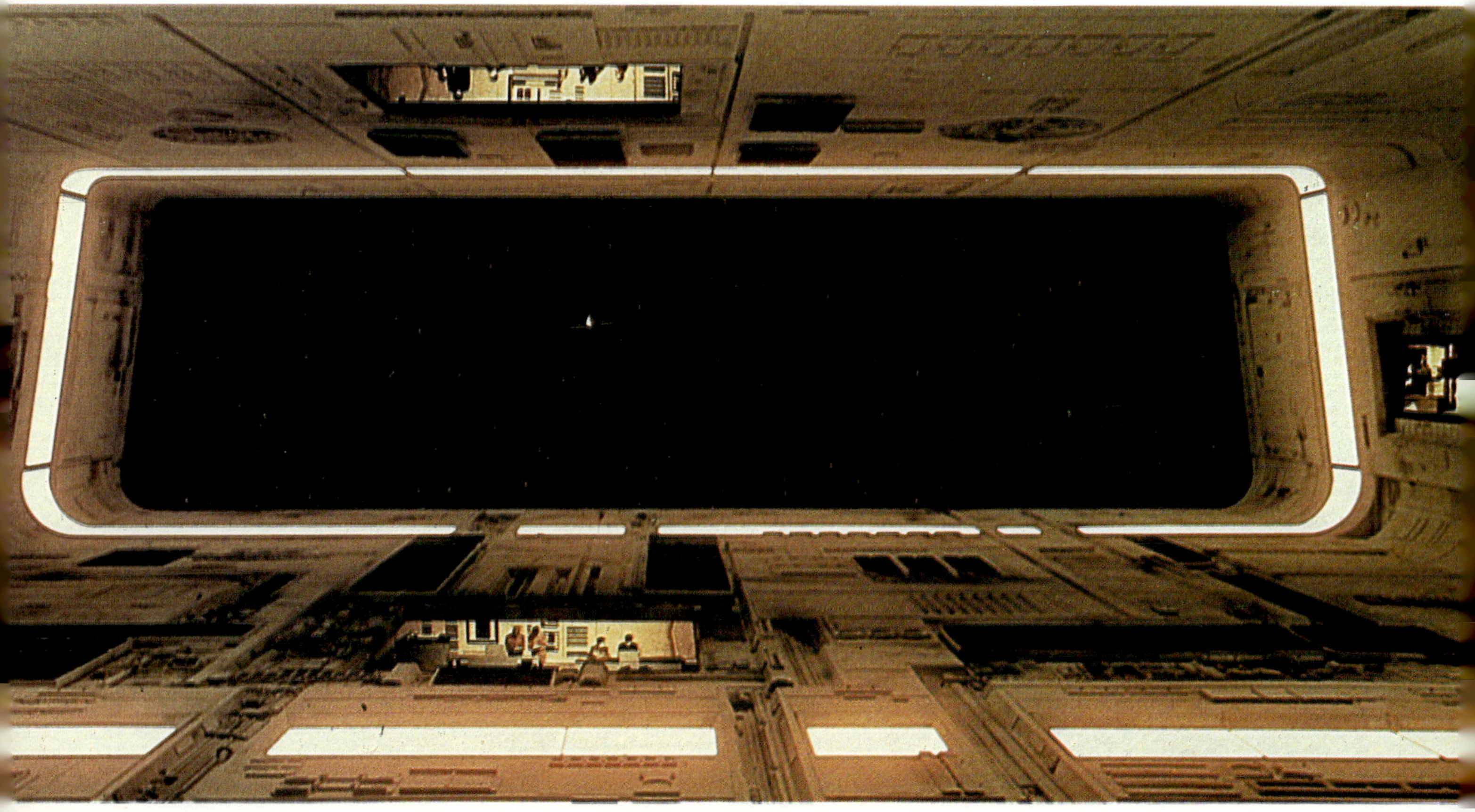

THE FUTURE OF THE FUTURE

What is coming next? Science fiction writers have not let us down in the past so we can look forward to reading in the future some really interesting speculation. In the fifties the more technically minded SF writers working from then-current knowledge attempted to visualize the future of the space programme. Now these first steps into space have been made and the space agencies themselves have taken over the role of preparing future scenarios. New spacecraft will indeed be built and we know the blueprints that they will follow. Any writer who ignores this will ensure that his story lacks all plausibility. This is all right for TV and the films where the scientific content usually borders on zero, but it won't do for science fiction. So once again the imaginations of the SF writers will have to range far into the future—with one difference. Unlike the writers of the thirties the writers today have some very sophisticated scientific concepts to draw upon, and this is reflected in their writing. Only alien craft from outer space, reflecting an alien science, can be utterly mysterious. This sort of story can lap over into fantasy, where the speculative writer of hard SF must stick closer to known facts. Still, either type of story might prove to be correct in the long run, as Arthur C. Clarke points out in *Profiles of the Future*:

Of the many lessons to be drawn from this slice of recent history, the one that I wish to emphasize is this. Anything that is theoretically possible will be achieved in practice, no matter what the technical difficulties, if it is desired greatly enough. It is no argument against any project to say: 'The idea's fantastic!' Most of the things that have happened in the last fifty years have been fantastic, and it is only by assuming that they will continue to be so that we have any hope of anticipating the future.

Clarke himself is partially responsible for a vision of the spaceship of the future that is almost becoming standard. In 1968, in collaboration with Stanley Kubrick, he wrote the screenplay for *2001: A Space Odyssey*. This was science fiction at its eye-banging best, and though there seemed some puzzlement as to just what the ending was all about, it can't be denied the technical effects were superb. As Ray Bradbury said, 'I think it's a gorgeous film. One of the most beautifully photographed pictures in the history of motion pictures.' What they were photographing were some of the most

Top left *A spaceship approaches the space station in* 2001. *Kubrick's film achieves effects still unmatched by any rival.*
Left *Looking out from the space station.*

116 THE FUTURE OF THE FUTURE

Above *The famous and much-imitated space train from 2001, dwarfing the small repair pod in the foreground.*
Right *Astronaut Keir Dullea talks to the computer Hal-9000 from inside the pod.*

THE FUTURE OF THE FUTURE

incredible models and effects ever created. It took eighteen months and $6,500,000 out of the budget to create this sense of reality and it was money well spent. The immense space train that took the Earthmen to the outer planets is still roaring across the covers of countless books. It set a pattern that a number of changes have been played upon.

One of the effects men from *2001*, Douglas Trumbull, went on to make *Silent Running*. Despite the fact that this film has a story line of stunning stupidity it nevertheless looks very good. The space greenhouses, right out of Tsiolkovsky, certainly have a sound feeling of reality.

Not only was an image of the future being fixed in the SF movies, the quality of the effects was set and improving. Buck Rogers's dangling spaceships with their trickling exhausts of smoke have fought their last battle. Audiences have now come to expect a certain quality and they are getting it. *2001* has had an effect on the film industry that still spreads in widening circles, like ripples from a stone thrown into a pond. John Dykstra had worked with Trumbull on the effects for *Silent Running,* and he set out on his own to do *Star Wars*. This film, the biggest money-maker of all time, depended on

The immense spaceships were the best features of Silent Running. *The dome houses a few precious plants rescued from a polluted Earth.*
Overleaf *Space battle from* Star Wars. *The Federation appears to be winning this round against the rebels.*

120 THE FUTURE OF THE FUTURE

Below *Inside the rebel headquarters the space fighters are prepared for their final assault on the Death Star.*
Right *Men gather reverently around the Mother Ship in* Close Encounters of the Third Kind *as it begins its musical messages.*
Below right *The Mother Ship descends in a blaze of incandescent light.*

THE FUTURE OF THE FUTURE 121

its effects. It took Dykstra and 75 assistants eight months to produce the 360 different effects shots—and the effort shows. Over 75 model spaceships were constructed for the film and a number of them were blown up with satisfactory results. The computer-controlled camera, a product of the latest technology, could recreate its movements exactly, giving a heightened illusion of reality to the battles in space. (In all of the effects done for *2001* the camera never moved.) So many and so detailed were the spacecraft that the producers of the film actually thought they could copyright the future. They began legal action against the makers of *Battlestar Galactica.* This action would be very funny if millions of dollars weren't involved. Only the lawyers will get rich, for an equally strong case could have been made against *Star Wars* for swiping every cliché in science fiction, and many times filming scenes and props actually taken from magazine covers. If there is any lawsuit against *Battlestar Galactica* it should be for prostitution of art.

The instant-classic picture of an alien spacecraft is now the magical-chandelier mother ship of *Close Encounters of the Third Kind.* This film is the culmination of a different tradition, not science fiction but UFOlogy. The desperate and yearning search for pie in the sky, manna from heaven. Heaven here is the floating city that

122 THE FUTURE OF THE FUTURE

Ralph McQuarrie's impression of a space battle in Battlestar Galactica, *another film whose effects are its only saving grace.*

promises salvation though, as with all mysticism, the promise is never quite spelled out. The philosophy of the film is as naïve as the effects are stunning. What it is all about is made clear in the ads which say: *we are not alone*. Seeing their wildest dreams acted out in glorious colour convinces the public that their hopes are true.

On television the image of the future is of course *Star Trek*. The immense popularity of the series has meant that the image of the *Enterprise* has been stamped on the retinas of old and young alike around the world. An enterprising American engineer drew up a set of blueprints for this spaceship that has since become a best seller; the only blueprints in history to have that distinction. Model kits have

THE FUTURE OF THE FUTURE 123

been sold in countless thousands, and a special sort of fan treasures them and anything else to do with *Star Trek*. These fans, called 'trekkies' by true science fiction fans and looked down upon as a lesser form of life, have a worldwide network of clubs, journals and conventions. 15,000 trekkies at a time will show up for one of these conventions, to hear the gospel from the lips of Captain Kirk and team, which is a phenomenon that bears close studying by the psychologists. The strength of this movement is of course shown by the American Shuttle which will finally carry the name *Enterprise*

Close up one can see that Battlestar Galactica's spaceships sport gun turrets which might have come straight from the belly of a World War Two Flying Fortress.

124 THE FUTURE OF THE FUTURE

Above *TV's most famous spaceship, the USS* Enterprise.
Left Fireball XL-5 *on the launching ramp.*
Above right *The square-jawed captain and robot co-pilot of the XL-5.*
Right Space 1999, *a TV series with some interesting spaceship designs, but no other redeeming features.*

THE FUTURE OF THE FUTURE 125

into space. Another tip of the hat is the British TV series *Blake's Seven,* which is a derivative of the original, even down to the TV clichés and resulting boredom. The *Enterprise* flies on, here named the *Liberator,* and is so important to the plot that it is actually one of the seven characters of the title.

Of course television budgets are not as limitless as those of the screen, so the model work leaves a certain amount to be desired. No one minded this in *Thunderbirds* and *Fireball XL-5* when the models were actually seen as models, strings and all, and flown by eye-bulging puppets. The team of Gerry and Sylvia Anderson did very well with these two series for children and they still look good upon re-run. But the weakness of the effects and the paucity of the plot became apparent when they produced *Space 1999.* The strings holding up the model spaceships are no longer visible; the ones supporting the plot are very evident. This series is purportedly science fiction, but is noted for its scientific inaccuracy.

There is our picture of the future, spread out before us. The flapping wings and sails of the past have given way to the jets and drives of the present. There will be changes in the future of spacecraft design, but they cannot be really radical ones. Too many artists and designers have spent too many years recently ringing changes on the craft that will hopefully rush through space one day as they soar through imagination now. Of course there will be new conceptions, but we are now at the pinnacle of design and there is only one more to come.

The real one, when they build it.

BIBLIOGRAPHY

ALDISS, B. *Science Fiction Art* (New English Library, London 1975; Crown Publishers, New York 1975)
BAILEY, J. O. *Pilgrims Through Space and Time* (Greenwood Press, Westport, CT 1972)
BAINBRIDGE, W. S. *The Spaceflight Revolution* (John Wiley & Sons, Chichester 1977; John Wiley, New York 1977)
BERNAL, J. D. *The World, The Flesh and the Devil* (Routledge & Kegan Paul, London & Boston 1970)
BERRY, A. *The Next Ten Thousand Years* (Jonathan Cape, London 1974; Saturday Review Press, New York 1974)
The Iron Sun (Jonathan Cape, London 1977; Dutton, New York 1977)
BRAND, S. (Ed.) *Space Colonies* (Penguin Books, Harmondsworth 1977)
BROSNAN, J. *Future Tense, the Cinema of Science Fiction* (Macdonald & Jane's, London 1978; St. Martin's Press, New York 1979)
CLARKE, A. C. *Profiles of the Future* (Victor Gollancz, London 1974; Popular Library, New York 1977)
(Ed.) *The Coming of the Space Age* (Victor Gollancz, London 1967; Meredith Corp, Iowa 1967) *The Exploration of Space* (Temple Press, Feltham 1951; Harper & Row, New York 1951)
Man and Space (Time Life Books, London and New York 1965)
CLARKE, I. F. *The Pattern of Expectation 1644–2001* (Jonathan Cape, London 1979; Basic Books, New York 1979)
CLEATOR, P. E. *Into Space* (Allen & Unwin, London 1953)
FREEDMAN, R. *2000 Years of Space Travel* (Collins, London 1963; Saunders, Philadelphia, 1963)
FREWIN, A. *One Hundred Years of Science Fiction Illustration* (Jupiter Books, London 1974)
HARRISON, H. *Mechanismo* (Pierrot, London 1978; Reed Books, New Hampshire 1978)
HEPPENHEIMER, T. A. *Colonies in Space* (Warner Books, New York 1978)
HOLDSTOCK, R. (Ed.) *The Encyclopedia of Science Fiction* (Octopus Books, London 1978, New York 1979)
KEYHOE, D. E. *The Flying Saucers are Real* (Hutchinson, London 1950; Holt, Rinehart & Winston, New York 1950)
LEY, W. *Rockets, Missiles and Space Travel* (Viking Press, New York 1951)
MALONE, R. & SUARES, J. C. *Rocketship* (Harper & Row, London & New York 1977)
MARTIN, Dr A. R. (Ed.) *Project Daedalus—Final Report on the BIS Starship Study* (British Interplanetary Society, London 1978)
MOSKOWITZ, S. *Seekers of Tomorrow* (Classics of Science Fiction Series, Hyperion Press, Westport CT 1974)
Explorers of the Infinite (Classics of Science Fiction Series, Hyperion Press, Westport CT 1974)
Under the Moons of Mars (Holt, Rinehart & Winston, New York 1970)
NICOLSON, I. *The Road to the Stars* (Westbridge/David & Charles, Newton Abbot 1978)
NICOLSON, M. H. *Voyages to the Moon* (Macmillan & Co, New York 1975)
O'NEILL, G. *The High Frontier: Human Colonies in Space* (Jonathan Cape, London 1976; William Morrow & Co. Inc., New York 1977)
RYAN, C. (Ed.) *Man on the Moon* (Sidgwick & Jackson, London 1953; Crowell-Collier, New York 1953)
SMOLDERS, P. *Soviets in Space* (Lutterworth Press, Guildford 1973; Taplinger Publishing, New York 1974)
STOIKO, M. *Soviet Rocketry: First Decade of Achievement* (David & Charles, Newton Abbot 1973; Holt, Rinehart & Winston 1972)
TSIOLKOVSKY, K. E. *Beyond the Planet Earth* trans. K. Syers (Pergamon Press, Oxford 1960)
TUCK, D. H. *Encyclopedia of Science Fiction and Fantasy* vol. 1, 1974; vol. 2, 1977 (Advent Press, Chicago)
TURNHILL, R. *The Observer's Spaceflight Directory* (F. Warne & Co., London & New York 1978)
VON BRAUN, W. & ORDWAY, F. I. *History of Rocketry and Space Travel* rev. ed. (T. Y. Crowell, New York 1969)

PERIODICALS BIBLIOGRAPHY

HEINLEIN, R. 'Shooting "Destination Moon"' *Astounding SF* July 1950
OBERG, J. E. 'The Hidden History of the Soyuz Project' *Spaceflight* August to September 1975
'Russia meant to win the "Moon Race"' *Spaceflight* May 1975
'Korolev and Khrushchev and Sputnik' *Spaceflight* April 1978
VICK, C. P. 'The Soviet Super Boosters' *Spaceflight* December 1973 and March 1974
WHAITES, L. 'The Artist in Science Fiction: Christopher Foss' *Science Fiction Monthly* vol. 1, no. 2, 1974

ACKNOWLEDGMENTS

Ballantine Books: 97 (from A. C. Clarke *Rendezvous with Rama*, 1973). Bonestell Space Art: 54–55, 104. British Interplanetary Society: 28L, 53. Jim Burns (Young Artists): 112. Cinema International Corporation: 116 (MGM), 122 and 123 (© Universal Studios Inc.). Columbia Pictures Industries Inc. ©1977, 1978, 1979: 120T, 121. Tony Dalton: 55 (Paramount). David & Charles Ltd: 96, 102–3 (both from Iain Nicolson *The Road to the Stars* 1978. Illustrated by Andrew Farmer). Don Davis: 63, 86T. Deutsches Museum: 28R, 29. Chris Foss: 110–11. Robert Hunt Library: 48. ITC Entertainments ©: 124B, 125. Peter Jones/Solar Wind Ltd: 94. King Features Syndicate: 42. Kobal Collection: 40 (London Films), 114T (MGM), 117 (Universal), 120B (20th Century Fox), 124T (Paramount TV). George Locke Library (photography by A. Hornak): 7, 11, 14TL, 15, 27. Robert McCall: 109, 113. Mansell Collection: 10B. Ron Miller: 50–1 and inset (C. Bonestell), 52 (Rolf Klep). Chris Moore: 106–7 (from Alfred Bester *Extro*, pub. Magnum 1979). NASA: 2–3, 22, 23, 24–25, 49, 67, 72R, 73, 74, 77, 79, 80, 81, 82, 84, 86R, 87, 88–9, 90, 91, 92, 93. National Army Museum: 18T. National Film Archive/Stills Library: Endpapers (MGM), 12T, 12–13 (Columbia), 26–7 (Cinegate), 56 (Universal), 56–7 (Fox), 57 (Columbia), 114B (MGM). Novosti: 20, 46–7, 64, 65, 66, 68, 69, 72L, 76, 83, 85, 100, 101, 107. Photri: 75T, 78. Picturepoint: 70. Radio Times Hulton Picture Library: 8B, 16, 19T, 26. Tony Roberts (Young Artists): 106. Rockwell: 88L, 89B. A. Schomburg: 58, 62. Science Museum: 1. Smithsonian Institution: 4–5. R. Sternbach: 105. Twentieth Century Fox © 1977: 118–9. Ultimate Publishing: 32–3, 37R, 44B, 62. Boris Vallejo: 108. Alan Vince: 58–9 (© Syndication International). Frederick Warne Ltd; 75B (from R. Turnhill *Observer's Spaceflight Directory*, 1978). Tim White: 98–9. Chelsea House Publishers: 41.

Two publications of the British Interplanetary Society have been immensely helpful in the preparation of this book: *Spaceflight* and the *Journal* of the BIS. Any information on the BIS should be requested from the Executive Secretary, 12 Bessborough Gardens, London SW1V 2JJ. Other items consulted were in the collection of the Science Fiction Foundation at North East London Polytechnic. This is the largest collection in Britain of science fiction and associated factual material. Special thanks also to G. M. Webb for loan of unobtainable material, aid and technical advice.

INDEX

Page numbers in italics refer to illustrations

A
Across the Space Frontier 51, *52*
Across the Zodiac 13
Adamski, George 56
Air Wonder Stories 29
Aldiss, Brian 44
Aldrin, Buzz 73, *78*, 78
Amazing Stories 31, 42
American Institute of Aeronautics and Astronautics 29
American Rocket Society 29
Analog 38, 58
Anders, William 75
Anderson, Gerry 125
Anderson, Poul 62, 96
Anderson, Sylvia 125
Apollo spacecraft *70*, 71, 74–81, *74*, *75*, *78*, *79*, *80*, *81*
Ariosto, Ludovico 8
Armstrong, Neil *78*, 78, 79
Arnold, Kenneth 56
Ash, Fenton 14
Astounding Science Fiction 38
Astronautica Acta 95
Atlas rocket 67
Atterley, Joseph (see also Tucker, George) 10

B
Battlestar Galactica 121, *122*, *123*
Bergerac, Cyrano de 9
Bergey, Earle 57
Bernal, J. D. 44, 62, 89, 104
Bernal sphere *90*, 91
Berry, Adrian 108
Beyond the Planet Earth 14, 21–23
Birkenhead, Earl of 14
Black Hole 113
Black hole *106*, 107, 108
Blake's Seven 125
Blish, James 60, 61
Bonestell, Chesley 50, 54, 55, 58, *104*
Borman, Frank 73, 75
Brick Moon, The 13
British Interplanetary Society 29, 53, 100
Brown, Howard 30, 42, *43*
Buck Rogers 41
Burns, Jim *112*, 112
Burroughs, Edgar Rice 14
Bussard, Robert W. 95
Bykovsky, Valery 71
By Rocket to the Moon 26

C
Calkins, Dick *41*
Campbell Jr., John W. 38
Captive Universe 44
Carnell, John 29
Carr, Gerald 86
Cernan, Eugene 82

Cities in Flight 60
Clarke, Arthur C. 23, 29, 52, 53, 62, 64, 95, 115
Cleator, P. E. 29
Close Encounters of the Third Kind 120, *121*, 121
Clothier, Bob 58
Columbus of Space, A 14
Congreve, Colonel William 17
Cooper, Gordon 71
Copernicus 8
Crashing Suns 36
Cunningham, Walter 74

D
Daedalus Project 100, *102–103*, 104
Dan Dare 58, 60
Davis, Don 63, 113
Day the Earth Stood Still, The 56, *56*
Destination Moon 54, 55
Discourse Concerning a New World, A 9
Disney, Walt 113
Dold, Elliott *37*, *38*, 42, 43
Dyson, Freeman 100

E
Eagle, The 58, *59*
Eagle, lunar module *70*, 78
Earth vs the Flying Saucers 57
Echo 1 Satellite 62
Edison's Conquest of Mars 14, 23
Edwards, Gawain 29
Eisele, Donn 74
Eyraud, Achille 13
Exploration of Space, The 52, 53
Explorer 1 satellite 50

F
Fagg, Ken 58
Fantasy and Science Fiction 58
Feoktistov, Konstantin 71
Fireball XL-5 *124*, 125
First Men in the Moon 10, *12*, 13
Flash Gordon 41
Flash Gordon Conquers the Universe 42
Flash Gordon's Trip to Mars 42
Flying saucers 56, 57
Forever War, The 108
Forward, Robert 105
Foss, Chris *110*, 110–112
Frau im Mond 26, *26*
From the Earth to the Moon 10, *10*, *11*, 13

G
Gagarin, Yuri *68*, 68, 69, 77
Galactic Matter and Interstellar Flight 95
Galileo 8
Ganswindt, Hermann 19, 20, 97
Garby, Mrs Lee Hawkins 31
Gateway 108, *109*
Gemini spacecraft *72*, 72, *73*, 73
Gernsback, Hugo 31
Gibson, Edward 86

Gift from Earth, A 97
Girl in the Moon, The 26
Glenn, John 71
Glushko, Vladimir 65
Golden the Ship Was—Oh! Oh! Oh! 64
Goddard, Robert H. 18, *22*, 23–25
Godwin, Bishop Francis 9
Greg, Percy 13
Griffith, George 13, 14
Grissom, Virgil 72

H
Haldeman, Joe 108
Hale, Edward Everett 13
Hamilton, Edmond 36
Hampson, Frank 58, 60
Harbou, Thea von 26
Harrison, Harry 44, 86
Heinlein, Robert 44, 53, 55
Honeymoon in Space, A *13*, 14

I
Icaromenippus 7
If 58
Imagination 58
Iron Sun, The 108
Islands in the Sky 52
Islands of Space 38
It Came From Outer Space 55, 56

J
Jones, Peter *94*, 112
Jet-assisted take-off 24
Jupiter Theft, The 105

K
Kepler, Johannes 8, 9
Keyhoe, Donald 56
Klep, Rolf *52*
Komarov, Vladimir 71, 76
Korolev, Sergei P. 65, 66, *68*, 72
Khrushchev, N. 65, 66, 71, 72, 73, 76
Kubrick, Stanley 115

L
L-5 Society 90
Lagrange, Joseph 90
Lang, Fritz 26
Lasser, David 29
Last and First Men 41
Leiber, Fritz 106
Leonov, Alexei 71, *72*
Ley, Willy 26, 28, 51, 52
Life for the Stars, A 61
Lindbergh, Charles 24
Linebarger, Paul 61
Lovell, James 73, 75
Lucian of Samosata 7
Luna spacecraft 82, *83*
Lunar rover *82*
Lunokhod 82, *83*, 83

M
Mad Roland (see *Orlando Furioso*)
Maler, F. W. 14

Man in the Moone, The 9, 9
Man on the Moon, 50, 51, 52, 55
Man without a Country, The 13
Manning, Laurence 29
Mariner spacecraft 10, 62, 86
McCall, Robert *109*, *113*, 113
McKie, Angus 112
Méliès, George 12
Mendeleyev 19
Mercury spacecraft *67*, 67, 69, 71
Method of Attaining Extreme Altitudes, A 24
Metropolis 26
Modell B rocket 26
Moffitt, Donald 105
Moore, Chris 106, *106–107*
Morey, Leo 42, 44
Mote in God's Eye, The 105

N
Narrative of Arthur Gordon Pym 10
National Air and Space Administration 67
Nedbaĭlo, N. *100*
New Worlds 29, 58
Newton, Isaac 19
Nicolson, Iain 44
Nikolayev, Adrian 71
Niven, Larry 97, 105
Non-stop 44

O
Oberth, Hermann 18, 25–28, 55
O'Neill, Gerard 89, 90
Orion Project 97
Orlando Furioso 6
Orphans of the Sky 44

P
Pal, George 55
Paul, Frank R. *30*, *34*, 34, *38*, 42, 45
Pendray, G. Edward 29
Physics Today 90
Poe, Edgar Allan 10, 12
Poggensee, Karl *16*
Pogue, William 86
Pohl, Frederik 108, 109
Pournelle, Jerry 105
Pratt, Fletcher 29
Prelude to Space 53, 55
Prince of Space, The 44, 90
Princess of Mars, A 14
Priroda i lyudi 14
Profiles of the Future 115
Proton booster 73

Q
Quinn, Gerald 58

R
R-7 'Sapwood' rocket *65*, 65, 66, 73, 76
Rakete, Die 26
Raymond, Alex *42*
Redstone Rocket *67*, 67, 69
Rendezvous with Rama 95
Road to the Stars, The 44

Roberts, Tony *106*, 112
Rocket, The 26
Rocket into Interplanetary Space, The 25
Rocketship Galileo 53
Rogers, Hubert *39*, 42, 43
Round the Moon 6
Ruggieri, Claude 17
Russell, Eric Frank 29

S
Sailship *62*, 62, *63*, *86*, 105, *105*
Salyut 1 space station 83, *85*, 85
Salyut 3   85
'Sapwood' R-7 rocket *65*, 65, 66
Saturn rocket 21, 67, 71, 73, 74, *75*, 79, 80, 83
Schachner, Nat 29
Schirra, Walter 74
Schneeman, Charles 43
Schomburg, Alex 44, 57, 58, 62
Science Wonder Stories 29
Serviss, Garrett P. 14
Sheckley, Robert 64
Shepard, Alan 67, 69
Silent Running 117, 117
Skyfall 86
Skylab *84*, 85, 86
Skylark of Space, The 31–34, *32*, *33*, 42
Skylark Three 33, 34
Skylark of Valeron 34
Sloane, T. O'Connor 39
Smith, Cordwainer 61, 64
Smith, Edward Elmer ('Doc') 31
Smith, Edward Elmer 31
Smith, Malcolm 58, *60*
Smith, R. A. *53*
Sokolov, Andrei *100*, *101*
Solonevich *60*
Somnium 9
Son of the Stars 15
Soyuz spacecraft *76*, 76, *77*, 83, *85*, 85
Space 1999   *124*, 125
Space colony 90–93
Spacelab *88*, 88
Space shuttle, American 86–88, *88*, 89, 92
Specialist 64
Sputnik 1 satellite 49, *64*, 66
Sputnik II satellite *65*, 66
Stanford torus *91*, 91
Stapledon, Olaf 41, 79
Star Trek 122, 123, *124*
Star Wars 117, *118–119*, *120*, 121
Startling Stories 57
Sternbach, Rick *105*, 113
Sunjammer 62
Sunjammer (see Sailship)

T
Tau Zero 96
Temple, William E. 29
Tereshkova, Valentina 71
Things to Come 40, 41
This Island Earth 56

Thrilling Wonder Stories 57
Thunderbirds 125
Titan rocket 73
To Outlive Eternity 96
Toftoy, Holger 47
Trip to the Moon, A 12
True History 7
Trumbull, Douglas 117
Tsiolkovsky, Konstantin 14, 18–23, *20*, 62, 89
Tsander, F. 62
Tucker, George 10, 12
Tukhachevsky, Marshal M. N. 65
Tupolev 65
2001: A Space Odyssey 52, 113, *114*, 115, *116*, 117

U
Under the Moons of Mars 14
Unparalleled Adventures of One Hans Pfaall, The 10

V
V-2 rocket 26, *28*, 29, 47, *48*, 48
Valigursky, Ed *61*
Vallejo, Boris *108*
Van Allen, James A. *49*
Vanguard rocket *49*, 49
Verne, Jules 6, 10, 12, 13, 25, 31
Verein für Raumschiffart 26, 28, 29
Von Braun, Wernher 29, 47, 48, *49*, 50, 52, 65
Vne Zemli 21 (see also *Beyond the Planet Earth*)
Voskhod 71, 72
Vostok 1 rocket *46*, *68*, 68, *69*, 69, 71
Voyage à Vénus 13
Voyage dans la Lune 8, 9
Voyage to the Moon, A 10
Voyage to the Moone 8, 9
Voyager spaceprobe 87

W
Wanderer, The 106
War of the Worlds, The 13, 14, 23
Webb, Gerry 104
Webb, James 83
Weird Tales 34
Wells, H. G. 10, 13, 14, 31, 41
Wesso (see Wessolowski, Hans)
Wessolowski, Hans *36*, *37*, 42, 43
When Worlds Collide 55, 55
White, Ed 73
White, Tim *98–99*
Wilkins, Bishop John 9, 10
Williamson, Jack 44, 90
Woolley, Richard 47
World in 2030 A.D., The 14, 14
World, the Flesh and the Devil, The 44
Wunderwelten 14

Y
Yegorov, Boris, 71
Young, John 72

Z
Zond 77, 82